大廚做的*Pasta*
真美味

自序
Foreword

　　投身飲食行業，或許是命運早已注定。吃，在我家向來是非常重要的一環。不管多忙碌，媽媽從不會讓我們吃得馬虎。即使到現在，一家人的家常話題也總離不開美食。或許是自小的薰陶，小時候最愛將眼前的食物煞有其事地攪作一番，幻想自己是一名大廚師，正在調製一道超級美食，自我陶醉一番！

　　談到義大利菜，立即會讓人想起義大利麵條或Pasta——義大利烹調的代表菜。麵條在義大利人心中確實佔有很重要的地位，就像米飯在我們中國人心中的地位一樣，絕對不能缺少。如有留意新聞，一兩年前在義大利曾出現抗議麵條漲價太厲害的示威，由此可見麵條對義大利人的重要性！

　　我也特別喜愛烹製義大利麵條，因麵條的可塑性十分高，葷素皆宜。忙碌時可用簡單的材料，不消一會就能做成美味醬汁，加在麵條上就變成一道美食；而最重要的是，在吃完後總能為我帶來幸福滿足的感覺。如多花一點心思和時間，也可選用一些較特別，甚至珍貴的材料，製成讓朋友驚嘆的Pasta。Pasta的多姿多采，總是令人驚喜不斷！我在這本書中為大家提供的50多道食譜，既有傳統經典，為人熟悉的款式，也有一些融合了不同烹調特色的食譜，希望可為大家帶來更多烹調的靈感。

　　　　　　　　　　　　　　　　　　　　傅季馨

The quest for gastronomic pleasure has always been a priority in my family. Food, not only Chinese food but also food of various countries and regions makes up a good part of my childhood memories. I can still recall vividly how the whole family rolled up our sleeves and helped in the meticulous preparation of festive delicacies for Chinese New Year. It might be a hectic scene in the kitchen, but it was something I looked forward to every year. That was the start of my love affair with cooking.

Pasta is almost synonymous with Italian cuisine and it has always been among my favourite food. I love pasta for its versatility. When I am in a hurry, I need just a few minutes and some simple ingredients and I will have cooked a delicious sauce for a gratifying pasta meal. When I am in no hurry, I use special ingredients and I will have prepared a sumptuous pasta meal to the delight of my friends. In this book, I have put together some 50 pasta recipes. There are classic recipes, express recipes as well as lavish recipes. I very much hope that this book will be a source of inspiration to you in the preparation of your signature pasta dishes.

Margaret Fu

目錄 Contents

20

煮出…最好吃的義大利麵
Look Good Taste Good

清新蔬素 Pasta

滋味海鮮 Pasta

香濃肉類 Pasta

114 醬汁的製作 Sauce Base Makingg

認識 義大利麵

About Pasta

義大利麵的種類和烹調
Types and Cooking of pasta

義大利麵條可分為兩大類：乾麵條及新鮮麵條
There are 2 major types of pasta: dry and fresh

乾麵條　Dry pasta

優質的麵條都是用硬小麥粉製成，在定型及烘乾後出售。乾麵條可分為長條狀，例如：義大利長麵、扁麵、天使麵等；短麵條例如：通心粉、筆尖麵、螺旋麵等。在製作時麵條會用銅製定模型做成粗糙表面；或用細滑物料製定模型做成光滑表面。

Most good quality dry pasta is made with hard duram wheat, made into different shapes and sizes and sold to consumers in dry form. Mostly dry pasta can be differentiated into long pasta, e.g. spaghetti, linguine, capellini; and short pasta, e.g. macaroni, penne, fusilli. Some forms of dry pasta are found to have a rough surface, while some have a smooth surface. This is the result of the use of different piping mould: with the bronze producing rough surfaces and Teflon mould producing smooth surfaces.

新鮮麵條　Fresh pasta

製作麵條的主要材料是麵粉及液體。液體可以是蛋液以製蛋麵、墨魚汁及水分混和製作墨魚麵、或是菠菜汁製成菠菜麵等，新鮮麵條需要冷藏並盡快食用。

Most fresh pasta is made with flour and any kind of liquid, e.g. eggs for egg pasta; a mixture of squid ink and water for squid ink pasta; spinach juice for spinach pasta; etc. Fresh pasta needed to be stored in refrigerator and be consumed as quickly as possible.

如何烹調麵條？ How to cook pasta?

　　無論是新鮮或乾麵條，煮麵條時都是以一大鍋滾水，加入適量鹽以增加味道。一般參考用量是每2公升水分加入2湯匙細鹽。

　　除了大片麵皮如千層麵皮外，一般煮麵條時都不用加入油分。為了避免麵條黏在一起，放入麵條後必須要攪拌，特別是開始煮的前幾分鐘。

　　麵條煮熟後，盛起後可以立即放入醬汁內烹煮，並不須過冷水。所以，在烹調麵條前，應先將醬汁準備好。

　　No matter if it's dry pasta or fresh pasta, always cook in a big pot of boiling water with the addition of abundant salt to give flavour to the pasta. For reference, every 2 litres of water will require 2 tablespoons of fine salt.

　　Unless large flat piece of pasta, e.g. lasagna, is in cooking, there is no need to add oil to the pasta cooking water. To prevent the pasta from sticking together, make sure to stir the pasta from time to time, especially at the beginning of the cooking.

　　After the pasta is cooked, remove from the cooking liquid and add immediately to the sauce, it is not necessary to rinse cooked pasta under cold water. Instead, it is important to have the pasta sauce ready and be well heated in a pan before the cooking of pasta.

烹調時間 Cooking time

乾麵條 Dry pasta

　　麵條的包裝上會列明所需的烹調時間，可作參考。如麵條在煮熟後會再與醬汁同煮的話，最好是只煮至8分熟，盛起後加入醬汁後再煮至熟透，可使麵條吸收醬汁的味道。

　　Please check on the suggested cooking time as written in the packaging. This is the guideline to the approximate time of cooking. However, if the pasta is to be cooked with a sauce, it is recommended to cook the pasta only until 80% done, and leave the pasta to be cooked further in the sauce to absorb flavour than simply tossing together.

新鮮麵條 Fresh pasta

　　新鮮麵條所需的烹調時間比乾麵條要短，一般都是5分鐘之內就可以煮好。

　　Fresh pasta can be cooked in a much shorter time, usually for less than 5 minutes until fully cooked.

自製義大利蛋麵
Hand-made Egg Pasta

4 人 persons　　1 小時 hr

材料 Ingredients　（可製約500克麵糰 yields approx. 500 grams pasta）

義大利「00」麵粉300克，小麥粉50克，蛋黃3顆，全蛋3顆
300 grams Italian flour, type "00", 50 grams semolina flour, 3 egg yolks, 3 whole eggs

做法 Method

1
義大利麵粉及乾麥粉拌勻。
Stir the Italian and semolina flour together in a large mixing bowl.

2-1　2-2
將麵粉中間撥空，放進雞蛋和蛋黃。
Make a well in the middle, then add in egg and egg yolks.

3
慢慢與麵粉拌勻成麵糰。
Gradually mix with the flour to form a dough.

4
再搓揉數分鐘。
Knead for a few minutes.

5
麵糰封好，冷藏最少30分鐘。
Wrap well and leave to rest for at least 30 minutes in refrigerator.

麵糰取出，分割成數份。
Divide the dough into portions.

先將麵糰壓扁成0.5公分厚，然後放入製麵機輾壓均勻。
To thin out by pasta machine, press the dough down to 5-mm thick and pass through the machine at biggest gap.

將麵糰的兩邊向著中間對摺一次，把製麵機調到最寬度數將麵糰輾平。
Fold the dough into 3-folds and pass through the machine again.

再重複2次至麵皮成整齊長方形。如以手製，須用桿麵棍桿開至均勻的厚薄度。
Repeat for 2 more times until the dough is in uniformed rectangular shape. Then gradually thin out to desired thickness. If by hand, roll the dough out with a rolling pin to even thickness.

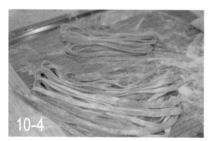

最後把麵皮用製麵機或刀切成所需形狀。均勻地灑上乾麥粉，防止麵條黏在一起，便可放冰箱儲存。

Using the pasta machine or by hand, cut the pasta into desired form. Sprinkle with a little semolina flour to prevent the pasta from sticking together. Cover and keep chilled.

Tips

1. 加入乾麥粉可增加麵糰口感。
2. 將麵糰輾開時，最好灑上較多乾麥粉以避免麵皮因黏着製麵機而遭扯破。

1. Addition of semolina flour gives more texture to the pasta.
2. Sprinkle the dough with generous amount of semolina flour to prevent it from sticking onto the pasta machine and thus breaking the pasta sheets.

材料 Ingredients

蛋麵皮約250克，黑松露醬1湯匙，鵝肝100克，濃雞高湯500毫升（參閱第117頁），切碎細香蔥1湯匙，蛋白1顆，打勻

Approx. 250 grams egg pasta dough, 1 tablespoon black truffle paste, 100 grams foie gras (goose liver), 500 ml brown chicken stock (Please refer to P.117), 1 tablespoon chopped chives, 1 egg white, beaten

做法 Method

1

鵝肝切小粒，與黑松露醬拌勻。
Dice foie gras and max with black truffle paste.

2

蛋麵皮壓成0.2公分的薄片。在桌面撒上少許乾粉，鋪上一片麵皮，刷上蛋白。
Roll the egg pasta dough out to 2-mm thin sheets. Line a pasta sheet on worktop and brush with egg white.

3

將鵝肝餡分成小份放在麵皮上。
Spoon portions of foie gras filling on the pasta sheet.

4

從麵皮的一端開始，將另一片麵皮鋪上。
Starting from one end, cover with a second sheet of pasta.

5

小心將每一份餡料的四周按實。
Press down tightly around the filling.

6-1

6-2

切割成單顆義大利餃。
Cut the ravioli into individual pieces.

15

材料 Ingredients

黃肉馬鈴薯500克，麵粉100克，細鹽少許，蛋液2湯匙
500 grams yellow flesh potatoes, 100 grams flour, a pinch of salt, 2 tablespoons beaten egg

做法 Method

馬鈴薯連皮煮熟，取出放涼，去皮並壓成薯泥。
Boil the potatoes with skin on until tender. Remove and leave to cool until warm. Peel and mash.

將薯泥、麵粉及細鹽混合。
Mix the mashed potato with salt and flour.

加入蛋液搓揉成麵糰。
Add beaten egg and form into a dough.

放檯上切成數片。

Knead on a table-top for a few times.

將麵糰切成小塊並搓成約1公分的粗繩狀。

Cut the dough into small pieces and roll each piece into a 1-cm thick rope.

切成2公分大的麵疙瘩。

Cut each rope into 2cm pieces.

麵疙瘩可保留平滑狀，或可將麵疙瘩用叉子背由上至下輕推，做出凹痕的效果。

Shape with a fork if desired by rolling each gnocchi on the back of a fork from top to bottom to form dents.

馬鈴薯麵疙瘩在盛盤前分別撒上適量麵粉，將麵疙瘩放在盤子上冷藏，使用時才取出。

Sprinkle the gnocchi and a tray with some dry flour. Place the gnocchi over and keep chilled or frozen until required.

Tips

製作馬鈴薯麵疙瘩最好選用紮實的馬鈴薯，如黃肉馬鈴薯。馬鈴薯於室溫下放置數天，使肉質更結實。

Choose firm flesh potatoes, e.g. yellow flesh potatoes for gnocchi making. For better results, leave the potatoes at room temperature for a few days to further firm up the flesh before cooking.

材料 Ingredients

義大利「00」麵粉200克，小麥粉50克，墨魚汁2茶匙，水約150毫升

200 grams Italian flour type"00", 50 grams semolina flour, 2 teaspoons squid ink, Approx. 150 ml water

做法 Method

墨魚汁及100毫升水拌勻。
Mix squid ink with 100 ml water.

兩種麵粉拌勻放入大碗內，中間留下一些空間，加入墨魚汁及一些水。
Stir together the 2 kinds of flour and make a well in the middle. Add liquid into the well with extra water.

與麵粉拌勻成麵糰；有需要可加入更多水。
Gradually mix with flour and form a dough.

麵糰放桌面搓揉至起筋。
Knead the dough a few times until elastic.

包好待30分鐘。
Wrap well and leave to rest for 30 minutes.

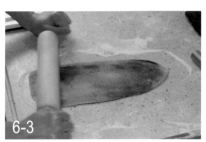

麵糰取出切成4-6塊，可用製麵機或用桿麵棍壓成所需的厚薄度。
Cut the dough into 4-6 pieces and thin out with pasta machine or by hand using a rolling pin to desired thickness.

麵皮撒少許小麥粉，放一旁待10分鐘至略乾。
Sprinkle the pasta sheets with extra semolina flour and leave aside to dry for 10 minutes.

麵皮捲起，切成所需形狀，放冰箱儲存。
Roll up and cut into desired shape by hand. Keep refrigerated.

煮出
最好吃的
義大利麵

Look Good Taste Good

蒜香橄欖義麵

Spaghetti All' olio, Aglio E Peperoncino

蒜香橄欖義麵

4 人 persons　　20 分鐘 mins

材料 | Ingredients

材料	Ingredients
義大利麵240克	240 grams spaghetti
蒜頭3粒，去皮、切碎	3 cloves garlic peeled & chopped
壓碎的乾紅辣椒¼茶匙	¼ teaspoon dried crushed red chili
初榨橄欖油4湯匙	4 tablespoons extra virgin olive oil
帕瑪森起士粉適量	Freshly grated Parmesan cheese to taste

做法 | Method

1. 將義大利麵煮熟。
2. 另於平底鍋內倒入4湯匙橄欖油並將油溫加高後，加入蒜頭用小火炒香使顏色呈現微黃，再加入壓碎的乾紅辣椒。
3. 拌入義大利麵（如太乾可加入少許煮麵水或熱水弄濕潤一點）。
4. 最後拌入帕瑪森起士粉及鹽、胡椒粉調味即完成。

1. Cook the spaghetti in salted boiling water until done.
2. In a frying pan, heat 4 tablespoons olive oil. Sauté the chopped garlic until fragrant but not brown. Add the dried crushed red chili.
3. Toss the spaghetti with garlic. If the pasta appears too dry, toss with a little pasta water.
4. Mix in grated Parmesan cheese and adjust seasoning with salt and pepper. Serve hot.

Tips

1. 此道是一款很簡單的義麵，只需選用優質材料，便可輕鬆做成美味可口的道地義大利麵條。橄欖油最好用初榨橄欖油，再配合鮮磨的正宗義大利帕瑪森起士，定會令人再三回味！
2. 初榨橄欖油不宜用太高溫烹調，否則香氣會消失。

1. Key to making this simple dish good is by using best quality ingredients. Extra virgin olive oil is recommended to give better flavour, and freshly grated Parmigiano Reggiano (Parmesan Cheese) will certainly enhance the flavour much more!
2. Do not overheat extra virgin olive oil, otherwise its flavour will be diminished.

茄醬天使細麵

Capellini with Salsa Pomodoro

材料
Ingredients

天使細麵240克 | 240 grams capellini (angel hair pasta)
蒜頭1粒 | 1 clove garlic
自製番茄醬200毫升（參閱第120頁）| 200 ml fresh tomato sauce (refer to p.120)
新鮮羅勒葉6-8片 | 6-8 leaves fresh basil
鹽、胡椒粉及糖各少許，調味 | Salt, pepper and sugar to taste
磨碎的帕瑪森起士 | Grated Parmesan cheese

做法
Method

1. 將3湯匙橄欖油的油溫加高，爆香蒜頭。
2. 加入自製番茄醬，用慢火煮開。
3. 天使細麵煮熟，取出瀝乾後加入熱番茄醬內拌勻，調味。
4. 將羅勒葉撕碎，加入細麵內拌勻，即可拌以帕瑪森起士食用。

1. Heat 3 tablespoons olive oil in a pan with crushed garlic clove.
2. Add tomato sauce and bring to a simmer.
3. Cook capellini in salted boiling water until al dente. Drain and add to the heated tomato sauce. Mix well with the sauce, season.
4. Hand shred fresh basil leaves and add to the pasta. Mix well and serve immediately. Serve with grated Parmesan cheese.

Tips

Fettuccine Alfredo 起源於羅馬一間餐廳，深受美國遊客歡迎，其後更廣傳於美國本土，成為美國人的其中一道經典義麵。最原始的做法，只需使用奶油及帕瑪森起士作醬汁，如想讓麵條更濕潤，可加入或改用鮮奶油代替鮮奶。這款義麵比較膩，所以加入適量肉荳蔻及檸檬汁，除了增加香味，更可提升食慾。

Originated from a restaurant in Rome, Fettuccine Alfredo was very popular among the American tourists. It was later brought to the U.S.A. and has become a classic pasta dish in the country. The original method for making Fettuccine Alfredo only requires the use of butter and grated Parmesan cheese. To make the pasta moister, cream or milk can be added.

奶油乳酪醬拌義式寬麵

 Fettuccine Alfredo

材料
Ingredients

義式寬麵320克	320 grams fettuccine
蒜頭1粒，切碎	1 clove garlic, minced
動物性鮮奶油100毫升	100 ml whipping cream
鮮奶100毫升	100 ml milk
凍奶油2湯匙	2 tablespoons chilled butter
即磨帕瑪森起士粉8湯匙	8 tablespoons freshly grated parmesan cheese
即磨肉豆蔻粉少許	A pinch of freshly grated nutmeg
檸檬汁1湯匙	1 tablespoon lemon juice

做法
Method

1. 義式寬麵用加鹽滾水煮至彈牙，盛起並留下一碗煮麵水。
2. 另將少許橄欖油的油溫加高並爆香切碎的蒜泥。加入鮮奶油及鮮奶，慢火煮5分鐘。用少許鹽及黑胡椒粉調味。
3. 加入義式寬麵及凍奶油拌勻。拌入起士粉、肉豆蔻及檸檬汁，再加入適量鹽、胡椒粉調味。如醬汁太稠，可加入適量的煮麵水。
4. 盛盤後趁熱食用。

1. Cook fettuccine in salted boiling water until al dente. Drain and keep aside 1 bowl of cooking liquid.
2. In a pan, heat drops of olive oil with minced garlic until aromatic. Add cream and milk, heat over low heat for 5 minutes. Check seasoning with a pinch of salt and black pepper.
3. Stir in cooked fettuccine and mix well. Stir in chilled butter and mix until the butter is melted. Stir in grated parmesan cheese, grated nutmeg and lemon juice. Check seasoning again. Add a little of pasta cooking liquid if the sauce is too thick.
4. Plate and serve hot immediately.

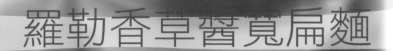

羅勒香草醬寬扁麵

Tagliatelle withPesto, New Potatoes &French Beans

羅勒香草醬寬扁麵

🥄 4 人 persons   ⏰ 30 分鐘 mins

材料 | Ingredients

寬扁麵240克   240 grams tagliatelle
馬鈴薯4顆   4 new potatoes
四季豆12條   12 French beans

羅勒香草醬 | Pesto Sauce

新鮮羅勒葉80克   80 grams fresh basil leaves
蒜頭1粒，去皮   1 clove garlic, peeled
松子仁或核桃仁20克   20 grams pine nuts or walnuts
帕瑪森起士粉50克   50 grams grated Parmesan cheese
鹽及黑胡椒粉調味   salt & black pepper to taste
初榨橄欖油約100毫升   appox. 100 ml extra virgin olive oil

做法 | Method

1. 羅勒葉洗淨擦乾，與其餘材料（除起士外）同放攪拌機內攪拌成泥狀，期間慢慢加入橄欖油，盛起備用。
2. 馬鈴薯與四季豆煮熟，然後把馬鈴薯切成小塊，放一旁保溫備用。
3. 麵條煮熟後，盛起後與羅勒香草醬、馬鈴薯、四季豆及帕瑪森起士碎拌勻，加入鹽及胡椒粉調味，即可馬上食用。

1. Clean and wipe dry the basil leaves. Combine all ingredients for the pesto, except grated Parmesan cheese, in a blender and blend until smooth by adding the extra virgin olive oil in a thin, slow stream. Remove and reserve.
2. Boil new potatoes and green beans until cooked. Cut the potatoes into quarters, keep warm.
3. Cook the tagliatelle in salted boiling water until done. Remove and toss with the pesto sauce, new potatoes, green beans and grated Parmesan cheese. Adjust seasoning with salt and pepper. Serve immediately.

Tips

1. 羅勒香草醬千萬不能直接加熱烹調，否則便會失去翠綠色澤及香味。
2. 儲存羅勒香草醬，最好放入乾淨容器內，然後以橄欖油蓋過醬汁以保存色澤。每次使用後，也要添加橄欖油蓋過餘下的羅勒醬。

1. Pesto sauce should never be cooked over direct heat or it will bruise and loses flavour.
2. To keep the pesto sauce, place in a clean container and cover with olive oil to retain the colour. Cover remaining sauce with extra olive oil after each use.

Tips

1. 要避免鮮奶油分解失去滑順感，應盡量避免用大火烹調太久。

2. Whipping cream 是一般最常見的動物性鮮奶油。如用脂肪含量較低的 single cream，可製成較稀釋的醬汁；如用脂肪含量較高的 double cream 則可做到一個更濃稠的醬汁。

3. 要維持滑順，則避免將帕瑪森風乾火腿烹調過長的時間。

1. Because of the higher fat content, cream is not recommended for boiling too hard. Otherwise, the cream will split.

2. Whipping cream is most easily found in HK; single cream can be used for a thinner sauce, or double cream for a thicker and creamier sauce.

3. Do not cook the ham for too long or it will be dried up.

帕瑪森起士奶油醬拌寬扁麵

Tagliatelle with Parma Ham, Peas and Parmesan Cream

材料
Ingredients

寬扁麵200克	200 grams tagliatelle
片裝帕瑪森風乾火腿150克，切絲	150 grams sliced parma ham, shredded
冷凍青豆4湯匙	4 tablespoons frozen garden peas
動物性鮮奶油100毫升	100 ml whipping cream
鮮磨帕瑪森起士粉4湯匙	4 tablespoons freshly grated Parmesan cheese
蒜頭1粒，切碎	1 clove garlic, minced
新鮮百里香1根	1 stalk fresh thyme

做法
Method

1. 青豆用鹽水汆燙1分鐘，撈起備用。
2. 寬扁麵煮熟。
3. 將1湯匙橄欖油的油溫加高，以慢火爆香蒜泥（小心不要讓蒜頭上色），加入鮮奶油、帕瑪森起士粉及百里香以慢火煮一會。
4. 麵條瀝乾，放入奶油醬汁內，加入青豆及火腿絲一起拌勻。調味後就可盛盤。

1. Blanch the garden peas in salted boiling water for 1 minute. Remove and drain.
2. Cook tagliatelle in salted boiling water until cooked.
3. Heat 1 tablespoon olive oil in a pan and sauté the minced garlic over low heat. Be careful not to brown the garlic. Add cream, Parmesan cheese and thyme, bring to a simmer.
4. Drain the pasta and add to the sauce. Toss well and add also the peas and parma ham. Adjust seasoning and serve hot.

Tips

1. Mac & Cheese 白醬起士焗烤通心麵相信是很多小朋友以及大朋友的最愛之一。黑松露給人高尚的感覺，味道濃郁，較適合大人口味，如為小朋友烹調，可選擇不用此醬。

2. 留意要預留較多白醬，另外醬汁不要太濃稠，因為通心麵在烤焗時會吸收更多水分。

1. Mac & Cheese is one of the well-loved dishes not only by young kids but by adult kids too. Black truffle is a prestigious kind of ingredient which will lend a sense of indulgence to the dish. Its taste is strong and is probably more suitable for adult's palate. If you are making this dish for children, you may omit the use of black truffle paste.

2. When preparing the béchamel sauce, you may want to make it thinner and make more of it. The macaroni will absorb more liquid during the baking process.

黑松露白醬起士焗烤通心麵

 Macaroni & Cheese

材料
Ingredients

通心麵240克	240 grams macaroni or mezzi rigatoni
蒜頭2粒，切碎	2 cloves garlic, minced
白醬1.5杯（360毫升）	360 ml (1.5 cups) bechamel sauce
黑松露醬1湯匙	1 tablespoon black truffle paste
鮮磨帕瑪森起士粉6湯匙	6 tablespoons freshly grated parmesan cheese
肉豆蔻粉少許	A pinch of grated nutmeg

調味
Seasoning

鹽、黑胡椒粉適量　　Salt & freshly ground black pepper to taste

做法
Method

1. 通心麵用已加鹽滾水煮至彈牙。撈起、瀝乾，並留下約1杯熱水備用。
2. 將1湯匙橄欖油的油溫加高，再用慢火爆香蒜泥，小心不要讓蒜泥上色。拌入白醬，慢火煮5分鐘至濃稠，然後拌入黑松露醬。
3. 拌入4湯匙起士粉及少許肉豆蔻粉。
4. 拌入通心麵，充分與醬汁混和；加入適量鹽、胡椒粉調味。
5. 將通心麵放入焗烤盤內，放入已預熱至攝氏200度烤箱內，烤至表面呈現金黃。趁熱食用。

1. Cook pasta in salted boiling water until al dente. Drain and reserve 1 cup of pasta cooking water on one side.
2. Heat 1 tablespoon olive oil in a saucepan with minced garlic. Be careful not to brown the garlic. Stir in bechamel sauce and cook over low heat for 5 minutes until the sauce is slightly thickened. Stir in black truffle paste.
3. Stir in 4 tablespoons parmesan cheese and grated nutmeg.
4. Stir cooked pasta into the cream sauce and mix well. Check seasoning with salt and black pepper.
5. Arrange the macaroni & cheese in a deep baking dish. Bake in preheated 200°c oven until golden brown on top. Serve hot immediately.

自製義式寬麵拌蘑菇奶油醬

Home-made Fettuccine with Mushroom Cream Sauce

自製義式寬麵拌蘑菇奶油醬

4 人 persons 25 分鐘 mins

材料 | Ingredients

材料	Ingredients
義式寬麵 300 克	300 grams fresh fettuccine
雞腿菇 2 朵，切片	2 drumstick mushrooms, sliced
白洋菇 8 粒，切片	8 white button mushrooms, sliced
乾牛肝菌 20 克，浸軟	20 grams dried porcini, soaked
油浸番茄乾 4 片	4 pieces sun-dried tomato in oil
蒜頭 2 粒，切碎	2 garlic cloves, minced
紅蔥 1 粒，切碎	1 shallot, chopped
乾百里香 ½ 茶匙	½ teaspoon dried thyme
乾馬沙拉酒 2 湯匙	2 tablespoons dry marsala wine
清雞高湯 50 毫升	50 ml chicken stock
動物性鮮奶油 200 毫升	200 ml cream
切碎的義大利巴西里 1 湯匙	1 tablespoon chopped Italian parsley

做法 | Method

1. 將 2 湯匙橄欖油的油溫加高，爆香紅蔥，加入所有的洋菇用大火一起炒，再加入切碎的蒜頭及灑上酒，收汁至差不多乾透時加入雞湯，用慢火煨煮 5 分鐘。
2. 加入番茄乾及動物性鮮奶油用慢火繼續煮 10 分鐘，調味。
3. 義式寬麵條煮熟，撈起，放入熱蘑菇奶油醬內，拌勻，調味後拌入切碎義大利巴西里就可趁熱食用。

1. Heat 2 tablespoons olive oil with chopped shallots. Add the mushrooms and fry over high heat. Add minced garlic and sprinkle with marsala wine, cook until dry. Add chicken stock and simmer for 5 minutes.
2. Add sun-dried tomatoes and cream, continue simmering for 10 minutes. Adjust seasoning.
3. Cook fettuccine in salted boiling water. Drain well and add to the hot mushrooms sauce. Toss well, adjust seasoning, sprinkle with chopped parsley and serve immediately.

Tips

1. 新鮮麵條一般在 5 分鐘之內便可煮熟，因此最好預先將醬汁準備好，再煮麵條。
2. 選擇新鮮蘑菇，以按下時可回彈且不下陷的為佳。

1. Fresh pasta can be cooked quickly within 5 minutes. Therefore, make sure the sauce is ready before boiling the pasta.
2. Choose firm mushrooms to get the freshest and best quality ones.

絲帶寬麵拌雞油菌菇、烤蒜及奶油起士醬

Pappardelle with Roasted Garlic & Mascarpone

材料
Ingredients

絲帶寬麵200克	200 grams pappardelle pasta
乾雞油菌菇4克，泡軟	4 grams dried chanterelle, soaked
雞腿菇50克，切細絲	50 grams fresh drumstick mushrooms, cut into strips
新鮮百里香1根	1 stalk fresh thyme
整顆蒜頭1個	1 head garlic
馬斯卡邦起士50克	50 grams mascarpone cheese
紅蔥1粒，切碎	1 shallot, minced
白酒50毫升	50 ml white wine
雞高湯100毫升	100 ml chicken stock

做法
Method

1. 蒜頭頂部切去小部份，灑上橄欖油並用鋁箔紙包裹。放入攝氏180度烤箱烤約45分鐘直到變軟。
2. 將2湯匙油的油溫加高，爆香紅蔥，加入雞油菌菇、雞腿菇及百里香用中高火同炒。拌入烤蒜醬及灑上白酒，煮至差不多乾透。拌入雞高湯，用慢火煮10分鐘。
3. 將絲帶寬麵煮至有嚼勁，拌入雞油菌菇內，加入馬斯卡邦起士拌勻，調味後趁熱食用。

1. Cut the top part of the garlic off. Drizzle with olive oil and wrap with a piece of aluminum foil. Roast in 180˚c oven for 45 minutes until very soft.
2. Heat a pan with 2 tablespoons olive oil. Sauté minced shallot, thyme, chanterelle and drumstick mushrooms over medium-high heat. Stir in roasted garlic paste and sprinkle with white wine, cook until almost dry. Add chicken stock and leave to simmer for 10 minutes.
3. Cook pappardelle until al dente. Add to the chanterelle sauce and toss well. Stir in marscarpone cheese and mix well. Check seasoning and serve hot.

核桃藍起士鮮奶油醬拌粗通心麵

Rigatoni with Blue Cheese Cream & Walnuts

材料
Ingredients

粗通心麵240克	240 grams rigatoni
藍起士（如Stilton或Gorgonzola）	50 grams stilton or gorgonzola,
50克，捏碎	crumbled
動物性鮮奶油200毫升	200 ml whipping cream
蒜頭4粒，去皮	4 cloves garlic, peeled
烘香核桃½杯	½ cup toasted walnuts
肉豆蔻粉少許	Pinch of grated nutmeg
鮮奶300毫升	300 ml milk
水300毫升	300 ml water
蜜糖少許	Honey

做法
Method

1. 將蒜頭用鮮奶和水各 100 毫升，慢火煮約 5 分鐘；瀝乾後再重複 2-3 次，至蒜頭變軟。
2. 鮮奶油與切碎的藍起士一起用慢火煮至滑順；蒜頭壓成泥，加入醬汁內拌勻，拌入肉豆蔻並調味。
3. 粗通心麵用加鹽的滾水煮熟，撈起後加入醬汁內拌勻。
4. 核桃略切碎，拌入通心麵內，調味後便可盛盤，灑上少許蜜糖即可趁熱食用。

1. Simmer the garlic cloves with 100 ml of milk and 100 ml of water for 5 minutes. Drain and repeat for 2 or 3 times until the garlic cloves are very soft.
2. Heat the whipping cream with crumbled cheese until smooth. Mash the garlic and add to the sauce. Stir in grated nutmeg and adjust seasoning.
3. Boil the pasta with salted boiling water until al dente. Drain and toss with the cheese sauce.
4. Coarsely chop the walnuts and add to the pasta. Adjust seasoning and plate. Drizzle with a little honey. Serve hot.

Tips

1. 茄子於正式烹調前先沾上麵粉後炸香，可助於保持形狀。如直接與醬汁同煮，很容易會煮得太軟。

2. 如用義大利進口茄子，最好先用鹽醃20分鐘，洗淨、擦乾後再裹粉炸香，便可去除茄子本身過多的水分及苦澀味。

1. Coating eggplants with flour and deep-frying can help to retain its shape. If leave to cook directly in sauce, the eggplants will become too soft and mushy.

2. If imported eggplants are used, leave to marinate with some salt for at least 20 minutes to remove its excess moisture and bitterness. Rinse well, dry, then coat with flour and deep-fry.

西西里茄子水管麵

 Rigatoni alla Norma

材料 / Ingredients

水管麵240克	240 grams rigatoni
本地茄子1隻	1 medium local eggplant
乾奧勒岡1茶匙	1 teaspoon dried oregano
蒜頭2粒	2 cloves garlic
紅葱1粒，切碎	1 shallot, minced
番茄醬200毫升	200 ml tomato sauce
切碎的乾辣椒少許	A pinch of chili
新鮮羅勒葉8-10片	8-10 leaves fresh basil
磨碎煙燻馬自瑞拉起士或	Smoked Mozzarella or grated Pecorino
Pecorino 羊乳酪適量	cheese to taste
麵粉適量	Flour for coating

做法 / Method

1. 茄子切成 1 公分小丁，以少許鹽調味後沾上麵粉。熱油炸至金黃，取出瀝油。
2. 將 2 湯匙橄欖油的油溫加高，爆香蒜頭及紅葱，拌入番茄醬、奧勒岡及切碎的辣椒用小火煮 5 分鐘。
3. 水管麵以加鹽的滾水煮熟至有嚼勁。
4. 將水管麵及茄子丁拌入茄汁內，撕碎羅勒葉後拌入水管麵中，接下來調味。趁熱拌入起士食用。

1. Dice eggplant into 1-cm cubes. Season with a little salt and coat with flour. Deep-fry until light brown, remove and drain.
2. Heat 2 tablespoons olive oil in a pan with crushed garlic and shallots. Stir in tomato sauce, oregano and chili. Leave to cook over low heat for 5 minutes.
3. Cook rigatoni in salted boiling water until al dente.
4. Add pasta and eggplant to the tomato sauce. Hand-shred the basil leaves and add to the pasta. Toss well and adjust seasoning. Serve hot with your choice of cheese.

Tips

1. 義大奶油乳酪遇熱便會融化，故在拌入南瓜泥前要先確定南瓜泥已經涼透才可。
2. 焦奶油會帶有一些果仁香味，與南瓜味道非常搭配。

1. Mascarpone cheese is a kind of Italian cream cheese with a light and soft texture. It will melt once being exposed to heat. Therefore, make sure to have the pumpkin pure?cooled down completely before stirring in the mascarpone cheese.
2. Burnt butter will deliver a nice nutty flavour which goes well with pumpkin.

南瓜義大利餛飩拌焦奶油

Pumpkin Tortelli with Burnt Butter

材料 Ingredients	蛋麵皮250克 帕瑪森起士100克，切碎 無鹽奶油4湯匙 蛋液，封口	250 grams fresh egg pasta 100 grams Parmesan cheese, grated 4 tablespoons unsalted butter Egg wash
南瓜餡 Pumpkin Filling	南瓜300克，去皮及切小丁 無鹽奶油20克 肉豆蔻粉少許 紅蔥2粒，切碎 馬斯卡邦起士2湯匙	300 grams pumpkin, peeled and diced 20 grams unsalted butter A pinch of grated nutmeg 2 shallots, chopped 2 tablespoons mascarpone cheese

做法
Method

1. 南瓜餡：紅蔥碎末用奶油炒軟，加入南瓜丁炒軟，有需要可蓋上鍋蓋，拌入肉豆蔻粉後壓成泥，待涼後拌入馬斯卡邦起士。
2. 將一片蛋麵皮放在桌上，每隔2公分放一湯匙南瓜餡。
3. 刷上蛋液。
4. 然後鋪上另一片蛋麵皮，小心沿餡料四周按實。
5. 分割成四方形餛飩。
6. 在餛飩上撒少許小麥粉，排在盤上後，保持冷藏。
7. 餛飩煮熟，撈起後瀝乾水分排放在熱盤上。
8. 將4湯匙奶油煮成焦色，刷在餛飩上並灑上帕瑪森起士粉，趁熱食用。

1. To prepare the pumpkin filling, sauté chopped shallot with butter until soft. Add pumpkin slices, cover with a lid and cook until soft. Remove from heat, add grated nutmeg and mash. Leave to cool. Stir in the mascarpone cheese and season well.
2. Line a piece of pasta on work table and place 1 tablespoonful of pumpkin filling on top, leaving 1-cm edge and 2-cm apart from each spoonful of filling.
3. Brush the pasta with egg wash.
4. Line another piece of pasta sheet over. Seal the edges around each spoonful of filling by pressing down tightly.
5. Cut each tortelli into square.
6. Sprinkle with more semolina and line on a tray. Keep chilled.
7. Cook tortelli in salted boiling water. Remove and drain well. Arrange on a hot plate.
8. In a pan, heat 4 tablespoons butter until brown. Drizzle over the tortelli and sprinkle with grated Parmesan cheese. Serve hot.

茄醬水管麵拌瑞可塔起士佐茄子及櫛瓜

Rigatoni with Eggplants & Zucchini in Tomato Sauce

材料
Ingredients

水管麵320克	320 grams rigatoni
瑞可塔起士100克	100 grams ricotta cheese
小茄子1根，切片	1 small eggplant, sliced
義大利櫛瓜1根，切片	1 zucchini, sliced
番茄醬200毫升	200 ml tomato sauce
乾奧勒岡1茶匙	1 teaspoon dried oregano
蒜頭2粒，拍碎	2 cloves garlic, crushed
紅葱2粒，切碎	2 shallots, chopped

做法
Method

1. 將 2 湯匙橄欖油的油溫加高，將茄子及櫛瓜片煎至兩面微黃，取出。
2. 將 1 湯匙橄欖油的油溫加高並爆香蒜頭。加入切碎的紅葱爆香，再加入番茄醬及奧勒岡，慢火煨煮 10 分鐘。
3. 水管麵用加鹽的滾水煮至有嚼勁，瀝乾後加入醬汁內。
4. 拌入茄子及櫛瓜片，加入瑞可塔起士拌勻。以適量鹽、胡椒粉調味，盛盤即完成。

1. Heat 2 tablespoons olive oil in a pan. Fry eggplants and zucchini slices until light brown. Remove.
2. Heat 1 tablespoon olive oil in a pan with crushed garlic. Add shallots and sauté until soft. Add tomato sauce and dried oregano. Simmer for 10 minutes.
3. Cook rigatoni in salted boiling water until al dente. Drain and add to the sauce.
4. Add fried eggplants & zucchini to the pasta. Add ricotta cheese and toss well. Check seasoning with salt and pepper. Serve hot.

馬鈴薯麵疙瘩拌黑松露佐南瓜醬

 ## Gnocchi with Black Truffle Paste & Pumpkin Sauce

材料
Ingredients

馬鈴薯麵疙瘩280克	280 grams potato gnocchi
南瓜200克，去皮及籽	200 grams pumpkin, peeled & deseeded
蒜頭2粒，切碎	2 cloves garlic, minced
鮮奶200毫升	200 ml milk
融化的奶油2湯匙	2 tablespoons melted butter
肉豆蔻粉1/8茶匙	1/8 teaspoon grated nutmeg
肉桂粉1/4 茶匙	1/4 teaspoon ground cinnamon
黑松露醬1茶匙	1 teaspoon black truffle paste

做法
Method

1. 南瓜切小丁。將2湯匙奶油加熱，將南瓜用慢火炒至金黃。加入蒜泥及鮮奶，蓋上鍋蓋後，以慢火將南瓜煮軟。
2. 南瓜攪拌成泥狀，再拌入肉豆蔻及肉桂粉。
3. 馬鈴薯麵疙瘩用加鹽的滾水煮至浮上水面，取出後加入南瓜泥內，加入黑松露醬，拌勻。以適量鹽、胡椒粉調味後即完成。

1. Cut pumpkin into small pieces. Heat 2 tablespoons butter in a pan and fry the pumpkin until golden brown over low heat. Add minced garlic and milk. Put on a lid and cook over low heat until the pumpkins are soft.
2. Blend pumpkins into puree. Stir in nutmeg and cinnamon.
3. Cook potato gnocchi in salted boiling water until they float to the top. Drain and add to the pumpkin sauce. Add black truffle paste and toss well. Check seasoning with salt and pepper and serve immediately.

鮪魚茄汁義大利麵

Spaghetti with
Tuna in Tomato Sauce

鮪魚茄汁義大利麵

🥄 4 人 persons ⏰ 35 分鐘 mins

材料 | Ingredients

材料	Ingredients
義大利麵280克	280 grams spaghetti
罐頭鮪魚150克	150 grams canned tuna fish
罐頭去皮番茄200克，切粗塊	200 grams canned peel tomatoes, coarsely chopped
油浸鯷魚柳2條	2 anchovy fillets in oil
洋蔥半顆，切小丁	½ onion, finely chopped
蒜頭2粒，切碎	2 cloves garlic, minced
切碎的乾辣椒少許	A pinch of dried crushed chili
乾百里香½茶匙	½ teaspoon dried thyme

做法 | Method

1. 將2湯匙橄欖油的油溫加高，慢火將洋蔥丁炒軟，約5分鐘。
2. 加入鯷魚柳、蒜泥、切碎的乾辣椒以及百里香。加入番茄及100毫升水。慢火煮15分鐘。
3. 義大利麵用加鹽的滾水煮至有嚼勁。取出加入番茄醬內。
4. 鮪魚分成小塊加入義大利麵內。拌勻及適量鹽、胡椒粉調味。趁熱食用。

1. Heat 2 tablespoons olive oil in a pan and sauté onion over low heat until very soft, approx. 5 minutes.
2. Add anchovies, minced garlic, dried crushed chili and dried thyme. Add canned tomatoes and 100 ml water. Simmer for 15 minutes.
3. Cook spaghetti in salted boiling water until al dente. Drain and add to the sauce.
4. Break tuna fish into small chunks and add to the spaghetti. Toss well and check seasoning with salt and pepper. Serve hot.

Tips

罐裝鮪魚分別有用鹽水或油浸兩種，油浸的較香及嫩滑；鹽水浸泡的則較清淡，可依個人口味選擇。

Canned tuna can be found in brine soaked or oil soaked. The oil preserved tuna is stronger in flavour and smoother in texture, while the flavour of the brine soaked tuna is lighter. You may choose either one according to your own preference.

Tips

1. 淡菜多生長於低溫水域地帶，野生或養殖的品種均有。選購時，要選擇口緊閉及重量重的，以確保淡菜新鮮及高品質。

2. 番紅花，又稱西藏紅花，是全世界最昂貴的香料。市面可找到絲狀或粉狀，但一般以絲狀的品質較好，只需要極少量即可。番紅花會為食物添加豐富的金黃色，其香味並不明顯，但可令菜色的口感更豐富。

1. Blue mussels are found in temperate or polar waters around the world. They are either wild or farmed. To buy mussels, choose ones that are tightly shut to make sure that they are alive.

2. Saffron is the most expensive spice on earth. It gives food a rich golden-yellow colour, and is characterized by a bitter and hay-like fragrance. Saffron is available in either strands or powder form, but with strands of better quality. To use saffron in cooking, usually a very samll pinch is enough.

番紅花淡菜奶油醬義麵

Spaghetti with Mussels in Saffron Cream Sauce

材料 Ingredients

義大利麵300克	300 grams spaghetti
新鮮藍淡菜20隻	20 pieces fresh blue mussels
義大利櫛瓜或翠玉瓜1小隻，切丁	1 small green zucchini or jade melon, diced
白葡萄酒150毫升	150 ml white wine
蒜頭1粒	1 clove garlic
紅葱1粒，切碎	1 shallot, finely chopped
新鮮百里香1根	1 sprig fresh thyme
番紅花絲少許	A pinch of saffron threads
動物性鮮奶油50毫升	50 ml cream

做法 Method

1. 將2湯匙橄欖油的油溫加高並爆香蒜頭，加入淡菜及撒上白酒，蓋上鍋蓋後用中大火煮至淡菜開口，將淡菜及醬汁分別盛起備用。
2. 另將2湯匙橄欖油的油溫加高，爆香紅葱，加入櫛瓜或翠玉瓜略炒，再加入淡菜醬汁、番紅花絲、百里香及鮮奶油以慢火煨煮一會。
3. 義大利麵煮至有嚼勁，撈起加入醬汁內，同時加入淡菜，煨煮一會至醬汁濃郁，調味後便可盛盤。

1. Heat a pan with 2 tablespoons olive oil and 1clove of crushed garlic until smoking hot. Add mussels with white wine, cover immediately and cook until the mussels are opened. Keep the mussels and juice apart.
2. Heat another 2 teaspoons olive oil and sauté the chopped shallots. Add diced zucchini or jade melon and sauté briefly. Add the mussel cooking juice, saffron threads, thyme and cream. Allow to simmer while the pasta is cooking.
3. Cook the spaghetti in salted boiling water until al dente. Drain, add to the sauce with the mussels. Leave to cook further to thicken the sauce. Adjust seasoning and serve immediately.

Tips

1. 如用義麵製作沙拉，小心不要將義麵煮過久至太軟，否則便完全失去嚼勁，最佳方法是參考義麵包裝上的烹調時間，煮熟後馬上過冷水，瀝乾水分，便可拌入醬汁。

2. 菲達起士是一種新鮮硬起士，味道鹹香，是涼拌沙拉的美味選擇。這款起士可於一般大型超級市場找得到，有時是罐裝用鹽水保存，又或用保鮮袋密封包裹販售。當菲達起士開封從鹽水取出後，應盡快食用以免變質。

1. To retain the al dente texture of the pasta, do not overcook it when making salad. The best way is to cook the pasta as suggested on the packaging. Once it's cooked, remove and rinse with cold water. Drain well and mix with the salad dressing.

2. Greek feta cheese is a fresh hard cheese with a savoury flavour. It is a culinary choice for making salad. Feta cheese can be found in large scale supermarket either in can with brine or in vacuum seal packed. Once the package is opened, it must be consumed quickly.

優格醬拌煙燻鮭魚螺旋麵

Smoked Salmon Fusilli Salad with Yoghurt Dressing

材料
Ingredients

螺旋麵240克	240 grams fusilli
煙燻鮭魚片100克	100 grams sliced smoked salmon
希臘菲達起士50克	50 grams feta cheese

優格醬 / Yoghurt dressing

原味優格4湯匙	4 tablespoons natural yoghurt
沙拉醬2湯匙	2 tablespoons mayonnaise
乾奧勒岡 ½ 茶匙	½ teaspoon dried oregano
蜜糖1湯匙	1 tablespoon honey
檸檬汁1湯匙	1 tablespoon lemon juice
鹽、黑胡椒粉適量，調味	Salt & black pepper to taste
芝麻葉或法國捲心菜，盤飾	Rocket or Frisee salad on side

做法
Method

1. 螺旋麵用加鹽的滾水煮熟。瀝乾後拌入 1 湯匙橄欖油，待涼。
2. 將優格醬的材料拌勻，備用。
3. 將煙燻鮭魚片切條，希臘菲達起士切丁。
4. 盛盤前將所有材料拌勻，並以適量鹽、胡椒粉調味。可在室溫或冷凍後食用。

1. Cook fusilli in salted boiling water until cooked. Drain and mix with 1 tablespoon olive oil. Leave to cool.
2. Mix all ingredients for yoghurt dressing together.
3. Cut smoked salmon into thick strips and break feta cheese into small dices.
4. Just before serving, mix all ingredients together and check seasoning with salt and pepper. Serve at room temperature or chilled.

Tips

1. 「Pasta Puttanesca」發源自義大利南部拿玻里，照字面解釋可能會有點不雅，因它的意思是指「妓女製作的義麵醬汁」。其名字來源眾說紛紜，但是不論它的由來，「Pasta Puttanesca」已成為今日義麵醬汁中最受歡迎及喜愛的其中一款。

2. Puttanesca醬汁所需的材料都是便於家中儲存的，如罐頭番茄、鯷魚柳、黑橄欖等，故隨時都可在家烹調。其用料也充份反映其南義風味，橄欖油、番茄等都是義大利南部的特產。

1. "Pasta Puttanesca" is originated from Naples, Southern Italy, literally meaning "Pasta cooked the whore way". There are many different saying on how the name is achieved, but none-the-less, pasta puttanesca is one of the most popular and best known pasta dishes around the world nowadays.

2. Ingredients for making this sauce can all be easily stored and therefore make it a very convenient dish to cook anytime at home. The use of olive oil, tomatoes, etc, are also a reflect of its origin from Southern Italy.

南義香草番茄汁螺旋麵

Fusilli with Puttanesca Sauce

材料
Ingredients

螺旋麵240克	240 grams fusilli
罐頭去皮番茄400克，切碎	400 grams canned whole peeled tomato, coarsely chopped
洋蔥 ½ 顆，切碎	½ onion, chopped
蒜頭2粒，去皮	2 cloves garlic, peeled
黑橄欖12粒，去核、切粗粒	12 black olives, pitted & chopped
鹽醃酸豆2湯匙，浸水及擠乾	2 tablespoons capers in salt, soaked and drained
鯷魚柳6條	6 anchovy fillets
乾奧勒岡1茶匙	1 teaspoon dried oregano
乾紅辣椒1隻	1 dried red chili

做法
Method

1. 先將2湯匙橄欖油的油溫加高並爆香蒜頭，加入洋蔥炒香，拌入切碎鯷魚柳略炒，再加入番茄及其醬汁一起煮1分鐘。
2. 加入其餘材料，用中慢火煮約10分鐘。
3. 螺旋麵煮熟，撈起放入醬汁內拌勻，以鹽、胡椒粉調味即完成。（可隨意灑上磨碎 Pecorino 羊乳酪或帕瑪森起士。）

1. Heat a pan with 2 tablespoons olive oil and a clove of garlic. Sauté onion, add chopped anchovy fillets. Stir in chopped tomato with its own juice and cook for 1 minute.
2. Add rest of the ingredients and leave to cook over medium-low heat for 10 minutes.
3. Cook fusilli in salted boiling water until al dente. Drain the pasta and add to the sauce. Toss together, adjust seasoning with salt and pepper and serve immediately. May serve with freshly grated Pecorino or Parmesan cheese.

辣茄醬蛤蜊肉燴菠菜義式寬麵

Spinach Fettuccine with Clams in Spicy Tomato Sauce

材料
Ingredients

菠菜義式寬麵300克	300 grams spinach fettuccine
新鮮蛤蜊600克（約1斤），	600 grams fresh clams,
浸鹽水及洗淨	soaked & rinsed
蒜頭2粒	2 cloves garlic
白葡萄酒100毫升	100 ml dry white wine
月桂葉2片	2 bay leaves

辣番茄汁
Spicy tomato sauce

番茄醬250毫升	250 ml tomato sauce
油浸鯷魚柳4條	4 anchovy fillets in oil
蒜頭1粒	1 clove garlic
朝天椒2隻	2 fresh chilies, chopped
乾奧勒岡1茶匙	1 teaspoon dried oregano
新鮮羅勒1束	1 stalk fresh basil

做法
Method

1. 將2湯匙橄欖油的油溫加高，並爆香2粒蒜頭，加入蛤蜊、月桂葉及白葡萄酒，蓋上鍋蓋煮至蛤蜊開口；熄火，備用。
2. 另將2湯匙橄欖油的油溫加高，再爆香1粒蒜頭，加入鯷魚柳、朝天椒、奧勒岡及羅勒莖，拌入番茄醬、蛤蜊湯汁及適量水分，小火煨煮最少15分鐘。
3. 菠菜義式寬麵煮至有嚼勁，撈起放入熱番茄醬汁內，煮至熟透，加入蛤蜊肉及調味。最後灑上少許初搾橄欖油、撒上羅勒葉碎片即可趁熱食用。

1. Heat 2 tablespoons olive oil in a saucepan with 2 garlic cloves until very hot. Add clams, bay leaves and white wine, cover with lid and cook until the clams are opened. Remove.
2. Heat another 2 tablespoons olive oil with 1 clove garlic. Add anchovy fillets, chilies, oregano and basil stalk. Add tomato sauce with clam juice and more water if necessary. Leave to simmer for at least 15 minutes.
3. Cook spinach fettuccine in salted boiling water until al dente. Remove and add to the hot sauce. Cook further and add cooked clams, adjust seasoning and finish with a touch of extra virgin olive oil. Sprinkle with chopped basil leaves. Serve hot.

龍蝦義式細扁麵

 Lobster Linguine

材料 Ingredients

長條細扁麵240克	240 grams linguine
龍蝦1隻	1 lobster
紅葱1粒，切碎	1 shallot, minced
蒜頭1粒，去皮	1 clove garlic, peeled
白蘭地酒2湯匙	2 tablespoons brandy
番茄醬50毫升	50 ml tomato sauce
鮮蝦高湯200毫升	200 ml shrimp stock
初榨橄欖油少許	Dashes of extra virgin olive oil
鹽、胡椒粉，調味	Salt and pepper to taste
切碎的義大利巴西里適量	Chopped Italian parsley for garnish

做法 Method

1. 龍蝦洗淨將肉抽取出來，切成大塊，備用。
2. 將 1 湯匙油的油溫加高，炒香紅葱，加入龍蝦肉用大火炒約 1 分鐘，撒上白蘭地酒，收汁至差不多乾透，以鹽、胡椒粉調味，盛起備用。
3. 於同一平底鍋內再加入 1 湯匙油，將油溫加高並爆香蒜頭，加入鮮蝦高湯及番茄醬，煮滾後轉慢火煮 15 分鐘。
4. 麵條煮至有嚼勁，撈起與龍蝦肉一起加入醬汁內再煮 5 分鐘，以鹽、胡椒粉調味。最後加入初榨橄欖油，並撒上切碎的義大利巴西里裝飾即可。

1. Clean and remove meat from the lobster, cut into chunks.
2. Heat 1 tablespoon olive oil and sauté the minced shallot until fragrant. Add the lobster meat, sauté for 1 minute over high heat. Sprinkle with brandy and cook until the liquid is almost dried. Season lightly with salt and pepper. Remove and keep aside.
3. In the same skillet, heat another tablespoon of olive oil with a clove of garlic. Add the shrimp stock and tomato sauce. Bring to a boil and then simmer for 15 minutes.
4. Cook the linguine in boiling salted water until al dente. Drain and add to the sauce with the lobster meat, cook for 5 more minutes. Season with salt and pepper, drizzle with extra virgin olive oil and sprinkle with chopped parsley.

Tips

　　義式米型麵是義大利麵的一種，雖然形狀似米飯，但其製作原料是小麥粉，製作方法亦與一般義大利麵相似，只是做成米形而已。義式米型麵口感細滑，適合用較多的醬汁製作，容易入口，是小朋友喜歡的選擇。

　　Orzo is a kind of pasta which looks like rice. The main ingredient for making orzo is durum wheat flour, and it is made the same way as of all kinds of dry pasta, and then shape into rice. Orzo has a smooth texture and it is best served with soupy sauce. This would make a favourite kind of pasta for children.

龍蝦湯義式米型麵拌紅蝦

Red Prawns in Lobster Bisque with Orzo

材料
Ingredients

義式米型麵240克	240 grams orzo
罐頭龍蝦湯200毫升	200 ml canned lobster bisque
清水150毫升	150 ml water
番茄醬50毫升	50 ml tomato sauce
洋葱1顆，切細粒	1 onion, finely chopped
蒜頭2粒，拍碎	2 cloves garlic, crushed
新鮮中型蝦16隻	16 medium-sized fresh red prawns
白蘭地2湯匙	2 tablespoons brandy
切碎的義大利巴西里或	1 tablespoon chopped Italian parsley or basil
羅勒1湯匙	

做法
Method

1. 中型蝦去頭，留著備用。蝦身去殼，保留完整或切粒。
2. 將1湯匙橄欖油的油溫加高，將蝦頭及蝦肉炒香，灑入酒並用少許鹽、胡椒粉調味。盛起備用。
3. 同一平底鍋內，將2湯匙橄欖油的油溫加高，並爆香蒜頭。加入洋葱用中火炒軟，拌入罐頭龍蝦湯、清水、番茄醬及蝦頭。煮滾後用慢火煨煮20分鐘，然後將蝦頭取出丟掉。
4. 義式米型麵以加鹽的滾水煮至有嚼勁，瀝乾後加入龍蝦湯內，繼續煨煮2分鐘，期間不停攪拌。加入蝦肉後以適量鹽、胡椒粉調味。
5. 將龍蝦湯義式米型麵盛盤，再撒上切碎巴西里或羅勒，趁熱食用。

1. Remove heads of the prawns and keep aside. Shell the prawns and keep them in whole or diced.
2. Heat 1 tablespoon olive oil in a pan and sauté the prawn meat and heads over high heat until just cooked. Sprinkle with brandy and season with a pinch of salt and pepper, remove.
3. In the same pan, heat 1 tablespoon olive oil with crushed garlic. Sauté chopped onion until soft over medium heat. Stir in canned soup, water, tomato sauce and sautéed prawn heads. Bring to a boil, then reduce heat and leave to simmer for 20 minutes. Remove and discard the prawn heads.
4. Cook orzo in salted boiling water until al dente. Drain and add to the lobster bisque. Keep stirring and cook further for 2 minutes. Add sautéed prawn meat and check seasoning with salt and pepper.
5. Plate and sprinkle each dish with chopped parsley or basil. Serve hot.

鮪魚義式米型麵沙拉
拌日式沙拉醬

Tuna & Orzo Salad
with Japanese Dressing

鮪魚義式米型麵沙拉拌日式沙拉醬

 4 人 persons　　25 分鐘 mins

材料	Ingredients
義式米型麵100克	100 grams orzo
鮪魚生魚片100克	100 grams sashimi grade tuna, diced
紅洋葱½隻，切碎	½ red onion, finely diced
番茄1隻，切碎	1 ripe tomato, diced
青紫蘇2片，切細絲	2 ohba leaves, finely shredded
烘香黑芝麻1茶匙	1 teaspoon toasted black sesame seeds

沙拉醬	Dressing
柚子醬油1湯匙	1 tablespoon yuzu soya sauce
味醂1湯匙	1 tablespoon mirin
日式米醋2湯匙	2 tablespoons Japanese rice vinegar
砂糖2湯匙	2 tablespoons sugar
麻油少許	Drops of sesame oil

做法　Method

1. 將日式米醋煮熱，拌入砂糖攪拌至熔化，再與其餘的沙拉醬材料拌勻。
2. 義式米型麵煮熟，與少許油拌勻，待涼。
3. 將所有材料拌勻，放涼後便可食用。

1. Stir sugar with heated rice vinegar until dissolved. Mix with other dressing ingredients and set aside.
2. Cook orzo until cooked through, drain and mix with some oil. Leave to cool.
3. Mix all ingredients together and keep chilled. Serve cool.

Tips

1. 柚子醬油是日式醬油的一種，具有日本柚子的香味。也可以用一般純味日式醬油加入新鮮檸檬皮或萊姆皮代替。
2. 味醂是甜味清酒，常用於日式烹調，可增加食物的甜味及光澤。

1. Yuzu soya sauce is a kind of Japanese soya sauce flavoured with citrus. Grated lemon or lime peel can be substituted with regular Japanese soya sauce.
2. Mirin is sweet sake, used widely in Japanese cooking for its sweetness.

Tips

1. 市場上的大貝殼麵許多都不用預先烹調。但是為了確保貝殼麵可完全被煮透，需要足夠水分，所以在烤焗中途如遇需要也可添加水分。

2. 如喜歡較濃起士味，可選用帕瑪森起士。

1. Most conchiglione available in the market do not require pre-cooking. But to cook the pasta thoroughly, enough amount of liquid is necessary. Add more water or chicken stock whenever the pasta appears too dry.

2. Parmesan cheese can be substituted if stronger cheese flavour is preferred.

白醬鮪魚焗烤貝殼麵

Baked Conchiglione with Tuna in White Sauce

材料
Ingredients

大貝殼麵200克	200 grams conchiglione
油浸罐頭鮪魚200克	200 grams canned tuna in olive oil
紅葱1粒，切碎	1 shallot, minced
蒜頭1粒，切碎	1 clove garlic, minced
鹽醃酸豆1湯匙，	1 tablespoon capers in salt, soaked
已浸泡及洗淨	& rinsed
白醬約300毫升（參閱115頁）	Approx. 300 ml béhamel sauce (refer to p.115)
清雞湯200毫升（參閱116頁）	200 ml chicken stock (refer to p.116)
馬自瑞拉起士50克，磨碎	50 grams grated Mozzarella cheese

做法
Method

1. 將 100 毫升白醬與起士拌勻，調味，放一旁備用。
2. 將 2 湯匙油的油溫加高，爆香紅葱及蒜泥，拌入鮪魚及酸豆快炒一會，再拌入白醬及清雞湯，煮滾後熄火並調味。
3. 在焗烤盤內刷上油。
4. 先將大貝殼麵吸飽鮪魚醬汁；將剩餘的醬汁鋪在焗烤盤底部，再排上大貝殼麵，最後以起士白醬覆蓋表面。
5. 蓋上鋁箔紙，放入已預熱至 180 度烤箱內烤 30 分鐘。除去鋁箔紙後再烤至表面呈現金黃色，取出趁熱食用。

1. Mix 100 ml béhamel sauce with grated mozzarella cheese, season and set aside.
2. Heat 2 tablespoons oil and sauté the minced shallot and garlic. Add tuna and coarsely chopped capers. Sauté briefly and stir in the remaining béhamel sauce and chicken stock. Bring to a boil, remove from heat and season with salt and pepper.
3. Brush a deep baking dish with oil.
4. Fill each conchiglione with tuna sauce. Line the bottom of the baking dish with the remaining tuna sauce and arrange the conchiglione over. Top with the reserved béhamel sauce and cover with an aluminum foil.
5. Bake in preheated oven at 180°c for 30 minutes. Remove the foil and continue baking until a golden crust is formed. Remove and serve hot.

鮮魷小番茄及酸豆拌蝴蝶麵
Farfalle with Squid
Cherry Tomatoes & Capers

鮮魷、小番茄及酸豆拌蝴蝶麵

🥄 4 人 persons ⏱ 25 分鐘 mins

材料	Ingredients
蝴蝶麵240克	240 grams farfalle
中型鮮魷2隻	2 medium size fresh squids
小番茄20粒，切半	20 cherry tomatoes, cut in halves
鹽醃酸豆1湯匙，泡水	1 tablespoon capers in salt, soaked
油浸鯷魚柳3條	3 anchovy fillets in oil
白酒100毫升	100 ml white wine
蒜頭1粒，去皮	1 clove garlic, peeled
紅葱1粒，切碎	1 shallot, finely chopped
魚高湯150毫升	150 ml fish stock
新鮮羅勒葉8-10片，切碎	8-10 leaves fresh basil, roughly chopped

做法

1. 將鮮魷洗淨，切成 1 公分粗圈。
2. 將 2 湯匙橄欖油的油溫加高，爆香蒜頭，加入紅葱炒香，再加入鯷魚柳及酸豆炒一會。
3. 加入魷魚圈，用大火快炒；灑入白酒後將醬汁煮至半乾。
4. 拌入小番茄及魚高湯（或以水將魚高湯塊煮溶）煮滾，用慢火煨煮10分鐘。
5. 將蝴蝶麵煮至有嚼勁，盛起加入醬汁內，拌勻後再煮數分鐘。調味後拌入羅勒葉，最後灑上初搾橄欖油即完成。

Method

1. Clean the squids and cut into 1-cm thick rings.
2. Heat 2 tablespoons olive oil in a pan with crushed garlic clove. Sauté the chopped shallot, add anchovy fillets and capers.
3. Add the squid and sauté quickly over high heat; sprinkle with white wine and let simmer until the wine is reduced by half.
4. Add cherry tomatoes and then fish stock. Bring to a boil and then leave to simmer for 10 minutes.
5. Cook farfalle in salted boiling water until al dente. Remove and add to the sauce. Toss well and allow to cook further for a few minutes. Adjust seasoning and add the chopped basil. Finish the pasta with a drizzle of extra virgin olive oil. Serve hot.

Tips

1. 魚高湯可用同等分量的清水，加入 1/2 魚高湯塊代替。
2. 魷魚必須用大火快速煮熟，以免出水太多。

1. Fish stock can be substituted with same amount of water + ½ cube of fish bouillon.
2. To prevent the squid from sweating too much , it is recommended to cook quickly over high heat.

Tips

1. 提起 PESTO 便會想起最常用的羅勒葉，其香味令很多愛美食的人士嚮往不已，但原來除了羅勒以外，其他新鮮的綠葉香草如巴西里、香菜等亦可用來製作香草醬，每一款都會有其獨特香氣。

2. 所有的 PESTO 香草醬、巴西里青醬都不適合烹調，最佳的食用方法是用來調配沙拉醬或將巴西里青醬拌入已煮好的義大利麵內，否則顏色會變黑，香味亦會遞減。

1. The most common ingredient to make pesto sauce is basil. Its lovely fragrance is highly praised by many gourmands. Apart from basil, other kinds of fresh green leaf herbs like parsley and coriander can also be used to make pesto. Each of these pesto sauces will depart its own unique flavour.

2. Just like the basil pesto, parsley pesto should not be heated up. The best way to make use of parsley pesto is salad dressing. Or it can also be mixed with cooked pasta to keep the bright green colour and its unique fragrance.

貝殼麵拌鮮蝦及巴西里醬

Conchiglie with Parsley Pesto & Prawns

材料
Ingredients

貝殼麵240克	240 grams conchiglie
鮮蝦16隻	16 fresh prawns
四季豆12條	12 stalks French beans
即磨帕瑪森起士粉4湯匙	4 tablespoons freshly grated parmesan cheese

巴西里青醬
Parsley Pesto

新鮮義大利巴西里葉1杯	1 cup loosely-packed flat leaf parsley
松子仁50克	50 grams pine nuts
初搾橄欖油120毫升	120 ml extra virgin olive oil
蒜頭1粒，拍碎	1 clove garlic, crushed

做法
Method

1. 所有巴西里青醬的材料攪拌成醬汁。放入乾淨容器內，再注入適量橄欖油覆蓋表面放在冰箱冷藏，最多可儲存4天。
2. 鮮蝦去殼，放入鹽水內煮熟，取出。
3. 四季豆去絲，放入鹽水內煮熟，取出。
4. 貝殼麵放入加鹽的滾水內煮熟，與鮮蝦及四季豆拌勻。
5. 拌入2-3湯匙巴西里青醬，拌勻並用適量鹽、胡椒粉調味。最後撒上帕瑪森起士粉即可。

1. Blend all parsley pesto ingredients to a purée. Keep in a clean container and cover with more olive oil on top. Keep chilled for up to 4 days.
2. Peel the prawns and blanch in salted boiling water until cooked. Remove.
3. Remove the ends of the French beans. Blanch in salted boiling water until cooked. Remove.
4. Cook pasta in salted boiling water until cooked. Remove and mix with the cooked prawns and French beans.
5. Stir in 2-3 tablespoons of parsley pesto and toss well. Season with salt and black pepper. Serve with grated parmesan cheese.

紅蝦、蠶豆、青花菜拌貓耳麵

Orecchiette with Red Shrimps, Fava Beans&Broccoli

材料
Ingredients

貓耳麵240克	240 grams orecchiette
紅蝦200克，去殼及挑腸	200 grams fresh red shrimps, shelled & deveined
冷凍蠶豆4湯匙	4 tablespoons frozen fava beans
青花菜¼個	¼ piece broccoli
鯷魚柳3條	3 anchovy fillets
蒜頭1粒，切片	1 clove garlic, thinly sliced
紅葱1粒，切碎	1 shallot, minced
乾辣椒1隻	1 dried chili
白酒100毫升	100 ml white wine
魚湯或水約150毫升	Approx. 150 ml fish stock or water

做法
Method

1. 青花菜切成小朵；蠶豆去殼。
2. 將青花菜煮到微熟；將蠶豆汆燙一分鐘，盛起。
3. 將2湯匙橄欖油的油溫加高，爆香蒜片、紅葱，加入鯷魚柳及辣椒乾，加入蝦隻炒至微熟，撒上白酒，待煮至半乾，加入蠶豆及青花菜炒勻，再加入魚湯用小火煮至濃郁。
4. 同時，貓耳麵用加鹽的滾水煮熟，撈起放入醬汁內，拌勻及略煮一會至水分收乾。調味後灑上初榨橄欖油，便可馬上食用。

1. Trim the broccoli and cut the florets into bite-size pieces; shell the fava beans.
2. Boil the broccoli until al dente; blanch the fava beans for 1 minute, remove and drain.
3. Heat 2 tablespoons olive oil in a pan with sliced garlic, shallot, anchovy fillets and dried chili. Add shrimps and sauté until just cooked. Sprinkle with white wine, reduce by half, add fava beans and broccoli. Add fish stock and leave to simmer.
4. Meanwhile, cook orecchiette in salted boiling water until cooked. Drain and add to the sauce. Toss well and leave to cook until the sauce is thickened. Adjust seasoning, drizzle with extra virgin olive oil and serve immediately.

清涼天使細麵
拌龍蝦及魚子醬
Chilled Capeliini with
Lobster & Caviar

清涼天使細麵
拌龍蝦及魚子醬

4 人 persons 25 分鐘 mins

材料	Ingredients
天使細麵200克	200 grams capellini (angel hair pasta)
龍蝦1隻	1 lobster
魚子醬2湯匙	2 tablespoons caviar
檸檬汁3湯匙	3 tablespoons fresh lemon juice
初搾橄欖油6湯匙	6 tablespoons extra virgin olive oil
切碎細香葱1湯匙	1 tablespoon chopped chives
鹽、胡椒粉，調味	Salt & pepper to taste
龍蝦油，隨喜好	lobster oil (optional)

做法

1. 龍蝦放入加鹽的滾水內煮熟，約 8-10 分鐘，取出過冷水後，抽取出龍蝦肉並切塊，備用。
2. 天使細麵煮至有嚼勁，撈起馬上用冰水過水，瀝乾。
3. 麵條與檸檬汁、橄欖油、龍蝦肉及 ½ 份魚子醬拌勻，調味後拌入細香葱。
4. 將麵條盛盤，再用少許魚子醬裝飾，也可灑上少許龍蝦油，涼拌吃。

Method

1. Boil lobster in salted boiling water until just cooked, about 8-10 minutes. Transfer to cold water. Remove the shell and dice the meat.
2. Cook the capellini until al dente, immediately transfer to ice water bath. Drain well.
3. Mix the capellini with lemon juice, olive oil, lobster meat and half of the caviar. Season with salt and pepper and sprinkle with chives.
4. Arrange on a serving plate and top with more caviar. Drizzle with lobster oil. Serve chilled.

Tips

1. 此道菜是於我曾工作過的 Gaia Ristorante 其中一道最受歡迎的經典前菜之一。特別感謝 Chef Paolo Monti 讓我登出其製作方法，與大家一起分享。
2. 如能用高品質的魚子醬，可增加此道義大利麵的美味；另外麵條不能過熟，否則便會有黏口的感覺。

1. This dish is an all-time hit at Gaia Ristorante, one of the best Italian restaurants in town. Special thanks to Chef Paolo Monti.
2. Key to success to this dish is the use of good quality caviar and do not overcook the capellini. The pasta has to be al dente in order to deliver a sense of freshness to the dish.

海膽燴特色長通心麵
Maccheroni with Sea Urchin Sauce

海膽燴特色長通心麵

4 人 persons   30 分鐘 mins

材料	Ingredients
長通心麵240克	240 grams maccheroni
新鮮海膽1塊	1 pack (large) fresh sea urchin
紅葱2粒，切碎	2 shallots, chopped
白葡萄酒100毫升	100 ml white wine
干貝2粒	2 dried conpoy

做法 | Method

1. 干貝浸 2 小時至軟，瀝乾後撕成細絲，另將干貝水留著備用。
2. 干貝絲與冷油一起放入鍋內，小火將干貝炸脆，撈起備用。
3. 另以炸干貝的用油 2 湯匙將紅葱用小火炒軟，加入白葡萄酒，煮至差不多完全蒸發，再加入干貝水 100 毫升用小火煮滾。
4. 麵條用加鹽的滾水煮至有嚼勁，盛起加入紅葱內，再煮一會。
5. 拌入海膽，拌勻後調味。
6. 將麵條盛盤後，撒上炸香的干貝即可趁熱食用。

1. Soak conpoy in water until soft, approx. 2 hours. Drain well and break into shreds. Reserve the soaking water.
2. Place the conpoy shreds in a pot and fill with oil. Heat slowly and fry the conpoy shreds until crispy. Remove and drain well.
3. In another pan, heat 2 tablespoons conpoy cooking oil and sauté the chopped shallots over low heat until soft. Add white wine and cook until the wine is almost evaporated. Add 100 ml of conpoy soaking water and bring to a simmer.
4. Cook the maccheroni in salted boiling water until al dente. Remove and add to the sautéed shallots. Cook further to desired doneness.
5. Add the sea urchin and toss well with the pasta. Adjust seasoning with salt and pepper.
6. Arrange the pasta in a plate and top with the fried conpoy. Serve hot.

Tips

1. 小心別用大火烹調海膽，否則便不能做到醬汁滑順的質感。
2. 要將干貝炸至鬆脆，最好用小火慢炸，直至油分內的泡泡消失。

1. Do not cook the sea urchin over very high heat or it will curdle and lose the silky texture.
2. To make sure that the conpoy is crispy, deep-fried over low heat until bubbles in the oil subside.

義大利扁麵拌蛤蜊白酒奶油醬

Linguine with Clams in White Wine Cream Sauce

材料
Ingredients

義大利扁麵280克	280 grams linguine
新鮮蛤蜊600克（約1斤）	600 grams fresh clams
蒜頭2粒	2 cloves garlic
乾朝天椒1隻	1 dried chili
白葡萄酒150毫升	150 ml dry white wine
動物性鮮奶油100毫升	100 ml whipping cream
義大利巴西里2湯匙，切碎	2 tablespoons finely chopped Italian parsley

做法
Method

1. 用寬邊平底鍋將橄欖油的油溫加高，並爆香蒜頭，加入蛤蜊、白葡萄酒及辣椒，立刻蓋上蓋子。用大火煮至蛤蜊開口，醬汁瀝乾後再放回鍋內，蛤蜊另留用。

2. 在蛤蜊汁內加入鮮奶油，用小火煮一會。

3. 同時間將麵條煮至有嚼勁，撈起瀝乾，放入蛤蜊汁內一起煮一會。加入蛤蜊拌勻，以鹽及胡椒粉調味，最後拌入切碎義大利巴西里，可即食用。

1. Heat a wide-base frying pan with olive oil and crushed garlic cloves until very hot. Add clams with white wine and chili, cover immediately. Cook over high heat until clams are opened. Reserve.

2. Heat clam juice in a saucepan and add whipping cream. Allow to simmer gently while the linguine is cooking.

3. Cook linguine in salted boiling water until al dente. Drain well and add to the clam sauce. Cook with the sauce for a while and add the clams. Adjust seasoning with salt and pepper and sprinkle with chopped Italian parsley. Serve immediately.

酥炸軟殼蟹拌檸檬奶油義式寬麵

Deep-fried Soft Shell Crab with Fettuccine in Lemon Butter Sauce

材料
Ingredients

義式寬麵240克	240 grams fettuccine
軟殼蟹4隻	4 soft shell crabs

炸漿
Deep-frying batter

天婦羅粉100克	100 grams tempura flour
冷水160毫升	160 ml cold water

醬汁
Pasta sauce

檸檬1隻，刨皮	Grated peel of 1 lemon
檸檬汁2湯匙	2 tablespoons fresh lemon juice
無鹽奶油50克	50 grams unsalted butter
動物性鮮奶油50毫升	50 ml whipping cream
紅葱1粒，切碎	1 shallot, finely chopped
切碎細香葱或義大利巴西里1湯匙	1 tablespoon chopped fresh chives or Italian parsley
鹽、胡椒粉，調味	Salt & pepper to taste

做法
Method

1. 天婦羅粉與冷水攪拌均勻成炸漿。
2. 軟殼蟹擦乾水分，沾上炸漿，放入熱油內炸至金黃香脆，取出用吸油紙吸去多餘油分。
3. 將奶油煮融，炒香紅葱，拌入檸檬汁。
4. 義式寬麵煮熟，與奶油一起加入醬汁內拌勻，調味後撒上細香葱或義大利巴西里。
5. 將麵條放盤子上，放上軟殼蟹，趁熱食用。

1. To prepare the batter, stir tempura flour with cold water until smooth.
2. Dry the soft shell crab probably. Dip into the tempura batter, deep-fry in hot oil until crispy and golden brown. Drain and leave on an absorbent paper to remove excess oil.
3. Melt butter in a frying pan and sauté the chopped shallots until soft. Add lemon juice.
4. Cook fettuccine in salted boiling water, add to the sauce with cream, toss well. Season with salt & pepper and sprinkle with chopped chives or Italian parsley.
5. Arrange the pasta on a plate and top with the fried soft-shell crabs. Serve hot.

金蒜蟹肉炒捲通心麵
Garganelli with
Crab & Golden Garlic

金蒜蟹肉炒捲通心麵

🥄 **4** 人 persons ⏰ **30** 分鐘 mins

材料	Ingredients
手捲通心管麵240克	240 grams garganelli
新鮮蟹肉200克	200 grams fresh crab meat
蒜頭2個，切碎	2 heads garlic, chopped
朝天椒4隻，切碎	4 bird's eye chili, chopped
鯷魚柳4條，切碎	4 anchovy fillets
紅葱2粒，切碎	2 shallots, minced
薑末1湯匙	1 tablespoon finely chopped ginger

做法 / Method

1. 蒜頭、紅葱及辣椒末一起與 200 毫升油放入鍋內，炒至微黃。
2. 加入薑末及鯷魚柳，繼續炒至金黃，盛起，瀝掉油。
3. 將少許蒜油的油溫加高，加入蟹肉。
4. 手捲通心管麵用加鹽的滾水煮熟，撈起加入蟹肉內，拌入金蒜，即可趁熱食用。

1. Place chopped garlic, shallots and chili in a pan with 200 ml oil. Fry until light golden brown.
2. Add ginger and anchovy fillets, allow to fry until golden brown. Remove and drain.
3. Heat some garlic oil in a saucepan and add crab meat.
4. Cook garganelli in salted boiling water until cooked. Drain and add to the crab meat with golden garlic, toss well and serve immediately.

Tips

1. 除手捲通心管麵外，任何短的筒狀義大利麵也適用。
2. 炒蒜泥時，注意油溫一開始時不能太熱，否則材料很快便會燒焦。整個過程最好是用小火慢慢炒香。

1. If garganelli cannot be found, any kinds of short pasta can be used in this recipe.
2. When cooking the chopped garlic, pay attention to the heat: only low heat should be used throughout the cooking process. Especially at the beginning, do not heat the oil until too hot, otherwise the ingredients will be burnt very quickly.

Tips

馬鈴薯麵疙瘩與烹調麵條一樣,都是需要大量滾水,放鹽,至浮水面即可,一般約需3-4分鐘。

To cook gnocchi, just like cooking pasta, requires a big pot of salted boiling water. Once the gnocchi started to float to the top, they are cooked and can be removed. This will usually takes 3-4 minutes.

義式馬鈴薯麵疙瘩拌蟹肉

Potato Gnocchi
with Crab Meat

材料
Ingredients

馬鈴薯麵疙瘩 400 克	400 grams gnocchi
（參閱第16頁）	(refer to p.16)

蟹肉醬
Crab meat sauce

蟹肉 200 克	200 grams crab meat
蒜頭 1 粒	1 clove garlic
罐頭去皮番茄 400 克	400 grams canned peel tomato
柳橙皮 ½ 顆，切碎	Peel from ½ orange, finely chopped
新鮮百里香 1 根	1 stalk fresh thyme
紅葱 2 粒，切碎	2 shallots, minced
動物性鮮奶油 2-3 湯匙	2-3 tablespoons whipping cream

做法
Method

1. 將 3 湯匙油的油溫加高，爆香蒜頭並將紅葱炒軟，加入番茄並壓碎，再加入 100 毫升水及百里香，用小火煮 20 分鐘。
2. 將蒜頭及百里香取出，加入柳橙皮碎末、蟹肉及奶油，以鹽、胡椒粉調味。
3. 馬鈴薯麵疙瘩煮熟，撈起加入熟蟹肉醬內再煮數分鐘。下調味及灑上初搾橄欖油後即可趁熱食用。

1. Heat 3 tablespoons oil with a clove of garlic and sauté the shallots until soft. Add canned tomato and crush. Add 100 ml water and thyme. Leave to simmer for 20 minutes.
2. Remove the garlic and thyme from the tomato sauce and add the orange peel. Add also the crab meat and cream; adjust seasoning with salt & pepper.
3. Boil the gnocchi with salted boiling water until cooked. Drain and immediately add to the crab meat sauce. Leave to cook for 5 more minutes and drizzle with a little extra virgin olive oil. Serve hot.

墨魚汁海鮮扁麵

Linguine with Squid Ink & Seafood Sauce

材料
Ingredients

義大利扁麵280克	280 grams linguine
墨魚汁1湯匙	1 tablespoon squid ink
魷魚1隻，切成圈狀	1 squid, cut into rings
蝦6隻，去殼及腸	6 prawns, shelled and de-veined
新鮮蛤蜊300克（約½斤），洗淨並浸鹽水	300 grams fresh clams, rinsed and soaked in salted water
扇貝4隻，切成4片	4 scallops, cut into quarters
蒜頭2粒	2 cloves garlic, peeled
白葡萄酒150毫升	150 ml dry white wine
鯷魚柳2條	2 anchovy fillets
番茄汁100毫升	100 ml tomato sauce
紅葱2粒，切碎	2 shallots, chopped finely
義大利巴西里適量，切碎	Chopped Italian parsley for garnish

做法
Method

1. 將橄欖油的油溫加高並爆香蒜頭，加入蛤蜊及灑上 100 毫升白葡萄酒，蓋上鍋蓋煮至蛤蜊開口後。熄火，將醬汁瀝乾備用。
2. 另將 2 湯匙油的油溫加高及爆香蒜頭，將魷魚、蝦及扇貝炒至半熟，灑入白葡萄酒及調味，盛起備用。
3. 再將 1 湯匙油的油溫加高，炒香紅葱及鯷魚柳，拌入番茄汁及蛤蜊汁同煮一會。
4. 同時地，將扁麵煮至有嚼勁，撈起放入醬汁內。
5. 海鮮全放回麵內及加入 1 湯匙墨魚汁拌勻，煮至醬汁濃郁，調味後灑上義大利巴西里碎末就可趁熱食用。

1. Heat 2 tablespoons olive oil with a clove of garlic until smoking hot. Add clams with 100ml white wine. Cover and cook until the clams are opened. Remove from heat, drain and reserve the juice.
2. Heat another 2 tablespoons olive oil with garlic. Sauté the squid, prawns and scallops until half cooked. Sprinkle with white wine and season with salt and pepper. Remove.
3. In the same skillet, heat 1 tablespoon oil and sauté the chopped shallots with anchovies until fragrant. Stir in tomato sauce and clam juice, leave to simmer.
4. Cook the linguine in salted boiling water until al dente. Drain and add to the sauce.
5. Return the seafood to the sauce and stir in 1 tablespoon squid ink. Toss well and cook until the sauce is thickened. Adjust seasoning and sprinkle with chopped Italian parsley. Serve hot.

1. 韭蔥的綠色部份的纖維較多，故多用作烹調高湯的材料。如要食用，應用白色部份。

2. 烹調韭蔥時，用小火可避免燒焦。

1. The green part of a leek is usually tough and strenuous. Therefore it is more suitable for making soup or stock. For cooking, only the white part is used.

2. Cook leek over low heat to prevent from getting burnt.quickly.

手打墨魚汁麵拌韭葱鮮蝦白酒醬

Squid-ink Pasta with Braised Leek & Prawn Sauce

材料
Ingredients

手打墨魚汁寬扁麵300克 （參閱第18頁）	300 grams hand-made squid ink tagliatelle (refer to p.18)

燴韭葱海鮮白酒醬
Braised Leek & Prawn Sauce

韭葱1根	1 leek
無鹽奶油20克	20 grams unsalted butter
中型蝦16隻，去殼及腸	16 medium-size prawns, shelled & deveined
白葡萄酒100毫升	100 ml white wine
蒜頭1粒	1 clove garlic
動物性鮮奶油100毫升	100 ml whipping cream
魚高湯300毫升	300 ml fish stock

做法
Method

1. 將2湯匙油的油溫加高並爆香1粒蒜頭，將中型蝦炒至半熟，灑入少許白葡萄酒後調味，盛起備用。
2. 將韭葱切碎，放入平底鍋內用奶油液以小火炒軟，灑入白葡萄酒50毫升煮至半乾。
3. 加入魚高湯煨煮15分鐘，加入奶油用小火煮至濃稠。
4. 墨魚汁麵條煮2分鐘。
5. 撈起即放入韭葱海鮮白酒醬內，加入中型蝦，拌勻再一起煮一下。用鹽、胡椒粉調味後即可盛盤，趁熱食用。

1. Heat 2 tablespoons oil with 1 clove of garlic. Fry the prawns until half cooked, sprinkle with some white wine and season with salt and pepper; remove and keep aside.
2. Chop the leek and sweat with melted butter. Add 50 ml white wine and reduce by half.
3. Add fish stock to the leek and braise for 15 minutes; add cream and cook over low heat for 5 minutes.
4. Cook the squid ink pasta in salted boiling water for 2 minutes.
5. Drain and add to the leek sauce. Add the prawns and toss well. Adjust seasoning and serve hot immediately.

Tips

美味優質的肉丸咬下時應有少許嚼勁，使口感更佳。使用新鮮的香腸便可得到這個效果，但留意要使用新鮮仍未煮熟的香腸，可於銷售進口食品較多的大型超級市場鮮肉櫃購買，如未能找到，可用絞肉回家自行調味及攪拌至有嚼勁。

When you bite into good quality meatballs, you'll find that these meatballs are quite elastic. By using ready-made sausage meat this can easily be achieved. However, only fresh uncooked sausage meat is applicable, which can be found at the fresh meat counter of large scale supermarkets selling imported goods. Fresh minced meat can be used in place of the sausage meat by seasoning and stirring to achieve the same quality at home.

小水管麵拌簡易肉丸汁

Bucatini with Quick Meatball Sauce

材料
Ingredients

小水管麵280克	280 grams bucatini
新鮮香腸200克	200 grams fresh sausage
番茄醬200毫升	200 ml tomato sauce
蒜頭1粒，切碎	1 clove garlic, minced
乾百里香½ 茶匙	½ teaspoon dried thyme
洋葱1顆，切碎	1 onion, finely chopped
動物性鮮奶油50毫升	50 ml fresh cream
磨碎的帕瑪森起士，拌食	Grated Parmesan cheese to taste
麵粉適量	Flour for coating

做法
Method

1. 香腸去腸衣，將絞肉捏成小丸子後薄薄地沾上一層麵粉。
2. 將鍋上的油溫加高，將肉丸煎至金黃，取出。
3. 鍋中約剩下 2 湯匙油，放入洋葱炒軟，加入蒜泥爆香，拌入番茄醬、百里香及 100 毫升水，小火煮滾後加入肉丸，蓋上蓋子煨煮 20 分鐘。
4. 小水管麵煮至有嚼勁，撈起，加入肉丸醬汁內，拌入動物性鮮奶油，一起煮一會兒至醬汁濃稠。調味後便可盛盤，與帕瑪森起士粉拌食。

1. Remove casing of the sausages and shape the minced meat into mini meatballs. Dredge with a thin coating of flour.
2. Heat a pan with oil and fry the mini meatballs until golden brown; drain.
3. Reserving 2 tablespoons oil in the pan, sauté the onion until soft. Add the minced garlic, stir well, then add tomato sauce, with thyme and 100 ml water. Bring to a simmer and add also the meatballs. Leave to cook for 20 minutes with lid on.
4. Meanwhile, cook bucatini in salted boiling water until al dente. Drain and add to the meatball sauce with cream. Cook until the sauce is thickened , season. Transfer to a platter and serve with grated Parmesan cheese.

Tips

1. 為避免蛋汁液與熱義大利麵接觸時凝固成塊狀，一定要先將蛋液與適量熱水混和。
2. 正宗 Spaghetti Carbonara 並不需要加入鮮奶油，主要是用微溫將蛋液煮至濃稠，從而做出滑順香濃口感。
3. 如用溫盤盛裝 Spaghetti Carbonara，效果會更佳。

1. To make sure that the eggs will not curdle when mix with hot pasta, it is very important to temper the eggs with some hot water first and then add to the pasta slowly.
2. Cream is not added to the authentic version of "Spaghetti Carbonara" The sauce is thickened only with beaten eggs cooked slowly to give the thick and silky texture.
3. Dish the Spaghetti Carbonara on warm plates will deliver an even better result.

蛋汁培根義麵

 ## Spaghetti Carbonara

材料
Ingredients

義大利麵240克	240 grams spaghetti
義大利培根100克，切小丁	100 grams pancetta, diced
雞蛋2顆	2 eggs
帕瑪森起士或Pecorino羊乳酪 50克，磨碎	50 grams Parmesan cheese or Pecorino cheese, grated
新鮮磨碎黑胡椒2茶匙	2 teaspoons freshly ground black pepper
動物性鮮奶油3湯匙（可選用）	3 tablespoons whipping cream (optional)

做法
Method

1. 將 1 湯匙橄欖油的油溫加高，加入培根及拌入較多的新鮮磨碎黑胡椒，用小火炒至金黃。
2. 將義大利麵用加鹽的滾水煮至熟透；取出加入培根內。
3. 將雞蛋打勻，拌入半份起士粉、動物性鮮奶油（如選用）及 100 毫升煮麵用的熱水。
4. 小心將蛋液慢慢拌入義大利麵內，迅速拌勻，繼續用小火將醬汁煮至濃稠，期間要不停攪拌。
5. 麵條盛盤，撒上其餘的起士粉就可以立即食用。

1. Cut the pancetta into small cubes and fry over low heat until golden brown with 1 tablespoon olive oil and abundant freshly ground black pepper.
2. Cook the spaghetti in salted boiling water until al dente. Remove and add to the fried pancetta.
3. Beat the eggs. Stir with half of the grated Parmesan cheese, cream if using, and 100 ml hot cooking water.
4. Add to the pasta slowly while stirring the pasta, stir well. Cook over very low heat and toss until the sauce thickens.
5. Dish up and sprinkle with the remaining grated cheese. Serve immediately.

Tips

1. 如想做成素食，不用加入培根粒，可按喜好加入2-3種蔬菜（各類蔬菜的甜味素合即可做成美味醬汁）。

2. 朝鮮薊在香港又稱雅枝竹，常用於地中海菜餚，其中心部份最美味。可於超市或雜貨店買到用橄欖油及醋醃的朝鮮薊心。

1. For vegetarians, simply omit the pancetta and add a couple more kinds of vegetables. The combination of sweetness from the vegetables will make a very nice pasta sauce too.

2. Artichoke is commonly used in Mediterranean cooking. The heart of the plant is considered to be the best part for eating. These can be found locally either frozen or preserved in olive oil and vinegar.

義大利培根春蔬燴筆尖麵

Penne with Pancetta & Spring Vegetables

材料
Ingredients

筆尖麵240克	240 grams penne
冷凍蠶豆100克，去皮及汆燙	100 grams frozen fava beans, peeled & blanched
蘆筍4條，切斜片	4 stalks green asparagus, sliced diagonally
冷凍朝鮮薊心4根，切成4塊	4 heads frozen artichoke hearts, quartered
義大利培根50克，切丁	50 grams pancetta in cubes
白葡萄酒100毫升	100 ml white wine
蒜頭1粒	1 clove garlic
蔬菜高湯或水200毫升	200 ml vegetable stock or water
月桂葉2片	2 bay leaves

做法
Method

1. 將培根丁以1湯匙橄欖油用慢火炒至金黃。
2. 加入拍碎的蒜頭爆香，加入朝鮮薊及蘆筍炒一會，撒入白葡萄酒，收汁至差不多乾透時，加入月桂葉、高湯或水煮滾後，改用慢火煨煮5分鐘。
3. 筆尖麵煮至有嚼勁，盛起後加入醬汁內一起煮一會。拌入蠶豆，下調味後灑上初搾橄欖油即完成。

1. Fry pancetta cubes with 1 tablespoon olive oil over low heat until golden brown.
2. Add crushed garlic, artichoke hearts and asparagus, sauté over high heat. Sprinkle with white wine and cook until almost dried. Add bay leaves, vegetable stock or water, boil and leave to simmer for 5 minutes.
3. Cook penne in salted boiling water until al dente. Drain and add to the sauce, leave to cook further. Toss in fava beans, adjust seasoning and drizzle with some extra virgin olive oil. Serve hot.

扇貝培根奶油醬拌義麵

Spaghetti with Bacon in Cream Sauce

材料
Ingredients

義大利麵240克	240 grams spaghetti
大扇貝8粒	8 large scallops
培根3片	3 slices bacon
紅葱1粒，切片	1 shallot, thinly sliced
蒜頭1粒，拍碎	1 clove garlic, crushed
乾奧勒岡 ½ 茶匙	½ teaspoon dried oregano
白葡萄酒50毫升	50 ml white wine
動物性鮮奶油100毫升	100 ml whipping cream

做法
Method

1. 培根切條，小火煎至金黃，將油分去掉，將培根取出備用。
2. 將 1 湯匙橄欖油的油溫加高，並爆香蒜頭，加入紅葱炒軟。加入白葡萄酒，小火煨煮至剩 2 湯匙的量。培根回鍋，加入鮮奶油及奧勒岡，小火一起煮 10 分鐘。
3. 扇貝用鹽及黑胡椒粉調味，放入熱鍋內用 1 湯匙油煎至兩面金黃，取出。
4. 義大利麵用加鹽的滾水煮至有嚼勁，取出麵條並留下 1 碗熱水備用。
5. 義大利麵加入鮮奶油醬汁內，煮至醬汁完全附著在麵條上。如醬汁太濃稠，可加入適量煮麵水稀釋。加入已煎香扇貝，以適量鹽、黑胡椒粉調味。趁熱食用。

1. Chop bacon coarsely and cook in a pan over low heat to render fat and until golden brown. Discard the fat and keep the bacon aside.
2. Heat 1 tablespoon olive oil in a pan with crushed garlic and sauté sliced shallot until soft. Add white wine and reduce to 2 tablespoons over low heat. Return bacon to the pan and add whipping cream and oregano. Simmer for 10 minutes.
3. Season scallops with salt and black pepper. Sear in a hot pan with 1 tablespoon oil until brown on both sides. Remove.
4. Cook spaghetti over salted boiling water until al dente. Drain and keep 1 bowl of cooking liquid aside.
5. Add cooked spaghetti to the cream sauce and cook until the pasta is coated with the cream sauce. Stir in a little cooking liquid if sauce appears too thick. Add seared scallops to the pasta and check seasoning with salt and pepper. Serve hot immediately.

香辣牛尾燴義麵

Spaghetti with Chili Oxtail Ragu

材料
Ingredients

義大利麵280克	280 grams spaghetti
牛尾800克	800 grams oxtail
胡蘿蔔½隻，切小丁	½ carrot, chopped
洋蔥1隻，切小丁	1 onion, chopped
西芹1根，切小丁	1 celery rib, finely chopped
罐頭番茄400克	400 grams canned tomato
濃縮番茄醬1湯匙	1 tablespoon tomato paste
紅葡萄酒200毫升	200 ml red wine
乾奧勒岡1茶匙	1 teaspoon dried oregano
月桂葉2片	2 bay leaves
乾朝天椒3-4隻	3-4 dried red chilies
新鮮朝天椒1-2隻	1-2 fresh chilies

做法
Method

1. 乾辣椒以乾鍋烘至微焦，備用。
2. 將牛尾調味，四周煎至金黃，取出，加入蔬菜炒香。
3. 拌入番茄醬炒到乾，放回牛尾，加入奧勒岡、月桂葉、番茄及乾辣椒，再加入紅葡萄酒，濃縮至剩下一半的量，然後加入足夠水分蓋過所有材料，煮滾後用小火煨煮2小時至熟透（如有需要，中途可再加入更多水分）。
4. 牛尾盛起，去掉骨頭，將肉放回醬汁內，加入朝天椒，再煮10-15分鐘。
5. 義大利麵煮至有彈性，撈起放入牛尾醬汁內拌勻調味即完成。

1. Toast the dried chili in a dry pan until charred. Remove.
2. Season and pan-fry the oxtail until brown on the outside, remove; add vegetables and sauté until soft.
3. Add tomato paste and fry until dry. Return the oxtail to the pot with dried oregano, toasted chili and bay leaves. Add red wine and reduce by half. Top up with enough water to cover all ingredients. Boil and then braise over low heat for 2 hours until tender. Add more water if necessary during the cooking process.
4. Drain the oxtails and remove the bone. Reheat the meat with sauce, add fresh chilies and cook further for 10-15 minutes.
5. Boil spaghetti in salted boiling water until al dente. Drain and add to the oxtail ragu. Toss well and adjust seasoning. Serve hot.

西班牙辣香腸、
黑橄欖、酸豆拌義麵

Rigatoni with Chorizo Sausages,
Olives & Capers

西班牙辣香腸、黑橄欖、酸豆拌義麵

4 人 persons　　25 分鐘 mins

材料　Ingredients

水管麵240克　　　　　　　240 grams rigatoni
西班牙辣香腸100克，切片　100 grams chorizo sausages, sliced
紅甜椒½隻，切細絲　　　　½ red bell-pepper, sliced
鹽醃酸豆2湯匙，洗淨及浸水　2 tablespoons capers in salt, rinsed & soaked
原粒黑橄欖12粒，去核及切碎　12 whole black olives, pitted & chopped
番茄醬200毫升　　　　　　200 ml tomato sauce
蒜頭2粒，切碎　　　　　　2 cloves garlic, chopped
乾奧勒岡½茶匙　　　　　　½ teaspoon dried oregano
義大利Pecorino羊乳酪　　　Grated Pecorino or
或帕瑪森起士粉適量　　　　Parmesan cheese to taste

做法　Method

1. 將1湯匙橄欖油的油溫加高，爆香蒜頭，加入酸豆用小火炒一會，再加入辣香腸及紅甜椒絲一起炒1分鐘，拌入番茄醬、奧勒岡及黑橄欖以小火煨煮10分鐘。
2. 水管麵煮至有嚼勁。
3. 水管麵撈起，放入醬汁內，拌勻及調味。最後拌入起士粉即完成。

1. Heat 1 tablespoon olive oil in a pan. Add garlic, then capers and fry over low heat. Add chorizo sausages and bell-pepper, fry for 1 minute. Add tomato sauce, oregano and black olives, leave to cook for 10 minutes.
2. Cook rigatoni in salted boiling water until al dente.
3. Drain the pasta and add to the sauce. Toss well and adjust seasoning. Serve with grated Pecorino or Parmesan cheese.

Tips

1. 要將黑橄欖去核，可用刀背下壓將橄欖壓裂，就可將核取出。
2. 西班牙辣香腸一般都是較油膩，故可不用加入太多其他油分烹調。

1. To remove pits from olives, crush olives with the back of a knife; pits can then easily be removed.
2. Chorizo is quite fatty and tends to extract oil during cooking. Therefore only a small amount of extra cooking oil is needed to cook the dish.

Tips

1. 義大利文「All' arrabiata」意即憤怒。此道義大利麵命名為「憤怒的義麵」，主要是因為加入了辣椒而成為一道辣勁十足的菜餚而來。
2. 一般都是選用「筆尖麵」Penne Rigate 去製作這道義麵，當然可隨意選用其他種類的義麵。

1. "All'arrabiata' means "angry style" in Italian. Name of this dish is derived from the addition of chili, which makes the dish hot and spicy.
2. Traditionally "enne rigate" is used for this dish. Of course, any other kinds of pasta can be substituted.

辣味番茄醬筆尖麵

 Penne All' Arrabiata

材料
Ingredients

筆尖麵240克	240 grams penne rigate
無骨雞腿肉150克，去皮	150 grams chicken thigh meat, skin removed
番茄醬200毫升	200 ml tomato sauce
乾朝天椒2-3隻	2-3 dried red chilies
蒜頭2粒，拍碎	2 cloves garlic, crushed
紅葱1粒，切碎	1 shallot, minced
乾奧勒岡 ½ 茶匙	½ teaspoon dried oregano
新鮮羅勒葉6-8片	6-8 leaves fresh basil
帕瑪森起士粉適量	Freshly grated Parmesan cheese to taste

做法
Method

1. 雞腿肉切成小塊及用鹽、胡椒粉調味。將1湯匙橄欖油的油溫加高並爆香1粒蒜頭，將雞肉用大火煎焦表面，盛起備用。
2. 將2湯匙橄欖油的油溫加高，爆香另1粒蒜頭及紅葱碎末，加入番茄醬、辣椒及奧勒岡，再加入適量水，用小火煮10分鐘。
3. 筆尖麵煮至有嚼勁，撈起放入辣味番茄醬汁內並加入雞肉，拌勻後調味及拌入切碎羅勒葉。盛盤與帕瑪森起士粉拌食。

1. Cut chicken meat into small pieces and season with salt and pepper. Heat 1 tablespoon olive oil in a pan with 1 crushed garlic clove, fry the chicken until well seared on the outside. Remove.
2. Heat 2 tablespoons olive oil with 1 garlic clove and minced shallot. Add tomato sauce, red chili and dried oregano. Add some water and leave to simmer for 10 minutes.
3. Cook penne in salted boiling water until al dente. Drain and add to the sauce with chicken meat. Toss well and adjust seasoning with salt and pepper. Sprinkle with chopped basil leaves and serve immediately with grated Parmesan cheese.

Tips

1. 風乾火腿的尾部因多筋的關係，故肉質較硬，不過因為較貼近骨骼，故味道也較濃郁，用作烹調最適合不過。如未能找到，用義大利培根也可代替。
2. 芝麻葉常用於沙拉，其帶胡椒辛辣的味道，也可用於義麵烹調，但是小心不要烹調過度，否則便會失去其獨特香味。

1. End part of a prosciutto (Parma ham) is usually a little chewy because of the sinew. However, it has a great flavour especially good for cooking. If prosciutto end cannot be found, pancetta can be substituted.
2. Rocket leaves, also known as arugula or rugula leaves, has a very special peppery flavour, which makes it not only great for salad but also for pasta cooking. However, it should not be cooked too much, otherwise flavour will be gone.

全麥筆尖麵拌風乾火腿

Prosciutto End with Whole Wheat Penne

材料
Ingredients

全麥筆尖麵240克	240 grams whole wheat penne
義大利風乾火腿尾部100克，切小丁或條狀	100 grams prosciutto end, cut into cubes or thin strips
乾百里香½茶匙	½ teaspoon dried thyme
紅葱1粒，切小片	1 shallot, thinly sliced
小番茄20粒，切成一半	20 pieces cherry tomatoes, halved
芝麻葉1小把，略切碎	1 small handful of rocket leaves, coarsely chopped
蒜頭1粒，拍碎	1 clove garlic, crushed
白葡萄酒100毫升	100 ml white wine
鮮磨黑胡椒粒1湯匙	1 tablespoon freshly ground black pepper
磨碎Asiago起士50克	50 grams grated Asiago cheese

做法
Method

1. 將3湯匙橄欖油的油溫加高，爆香蒜頭，加入火腿肉及黑胡椒粒，用慢火炒至微黃。
2. 加入紅葱爆香，拌入小番茄及百里香，撒入白葡萄酒，煮至半乾，再加入約100毫升水，小火煨煮一會。
3. 將全麥筆尖麵煮至彈牙，取出加入熱醬汁內拌勻（如太乾可加入更多煮麵熱水）。
4. 最後拌入芝麻葉及調味，盛盤後撒上Asiago起士粉即完成。

1. Heat 3 tablespoons olive oil with a clove of crushed garlic, add prosciutto with black pepper. Sauté over low heat until golden.
2. Add shallots, toss in cherry tomatoes and thyme. Sprinkle with white wine and reduce by half. Add approx. 100 ml water and leave to simmer.
3. Meanwhile, cook whole wheat penne in salted boiling water until al dente. Remove and add to the sauce. Toss well, add more pasta cooking water if the pasta appears too dry.
4. To finish, toss in rocket leaves, adjust seasoning. Transfer to a serving plate and sprinkle with grated Asiago cheese. Serve hot.

培根茄醬螺旋麵

Fusilli All'Amatriciana

培根茄醬螺旋麵

🥄 **4** 人 persons ⏰ **25** 分鐘 mins

材料	Ingredients
螺旋麵240克	240 grams fusilli
義大利培根80克，切成1公分小丁	80 grams pancetta, cut into 1 cm cubes
蒜頭1粒，拍碎	1 clove garlic, crushed
紅蔥1粒，切碎	1 shallot, minced
乾辣椒碎1茶匙	1 teaspoon dried crushed chili
乾百里香½茶匙	½ teaspoon dried thyme
乾奧勒岡½茶匙	½ teaspoon dried oregano
番茄醬200毫升	200 ml tomato sauce
義大利Pecorino羊乳酪粉3湯匙	3 tablespoons grated Pecorino cheese
帕瑪森起士粉3湯匙	3 tablespoons grated Parmesan cheese
義大利巴西里，切碎	Italian parsley, chopped

做法 | Method

1. 將1湯匙橄欖油的油溫加高，將培根丁用小火炒至金黃，盛起備用。
2. 爆香蒜頭，炒香紅蔥，加入番茄醬、香草、辣椒粉及適量水分。煮滾後轉小火煮15分鐘，放回培根丁。
3. 將螺旋麵煮至有嚼勁，撈起，放入醬汁內拌勻（如醬汁太濃稠可加入適量煮麵熱水）。拌入兩種起士粉，最後以鹽、胡椒粉調味，撒上巴西里碎末即完成。

1. In a saucepan, heat 1 tablespoon olive oil and sauté the pancetta cubes over low heat until golden brown. Remove and reserve.
2. In the same pan, sauté the garlic clove and minced shallot, add the herbs, tomato sauce, dried chili and some water. Bring to a boil and leave to simmer for 15 minutes; return the pancetta cubes to the sauce.
3. Boil the fusilli in salted boiling water until al dente. Drain and add to the tomato sauce. Toss well and add more pasta water if the sauce appears too thick. Sprinkle with grated Parmesan and Pecorino cheese, adjust seasoning and sprinkle with chopped Italian parsley. Serve immediately.

Tips

一般乾香草的味道會比較濃烈，所以使用時不用加入太多，否則很容易會蓋過其餘材料的味道。

Dried herbs have a very strong flavour, therefore, use with caution for cooking. If too much is being used, the dish will be over-powered by the flavour of the dried herbs.

迷迭香新鮮香腸
燴細寬蛋麵
Egg Tagliolini Rosemary
Scented Salsiccia

迷迭香新鮮香腸燴細寬蛋麵

4 人 persons　　**25** 分鐘 mins

材料	Ingredients
細寬蛋麵240克	240 grams egg tagliolini
Salsiccia 新鮮香腸200克	200 grams salsiccia or fresh sausage
新鮮迷迭香1根	1 sprig fresh rosemary
白葡萄酒100毫升	100 ml white wine
水100毫升	100 ml water
蒜頭1粒	1 garlic clove
洋葱 ½ 顆，切碎	½ onion, chopped
冰無鹽奶油20克	20 grams unsalted butter, chilled
帕瑪森起士粉50克	50 grams grated Parmesan cheese

做法 / Method

1. 香腸去腸衣並撕成小塊。
2. 將 2 湯匙油的油溫加高，爆香蒜頭，加入香腸碎塊煎香，取出。
3. 加入洋葱丁炒香，香腸回鍋，加入迷迭香、白葡萄酒及水，用慢火煨煮 15 分鐘。
4. 細寬蛋麵煮至有嚼勁，將水瀝乾後放入香腸醬汁內，拌勻後調味，最後拌入一小塊奶油，撒上起士粉便可食用。

1. Remove casing of salsiccia and break into small pieces.
2. Heat 2 tablespoons oil with garlic clove in a pan. Sauté the salsiccia until light brown on both sides. Set aside.
3. Add chopped onion and sauté until soft. Return the salsiccia and add chopped rosemary, white wine and water. Simmer for 15 minutes.
4. Boil egg tagliolini in salted boiling water until al dente. Drain and add to the salsiccia. Stir well and adjust seasoning. Stir in a knob of chilled butter and Parmesan cheese. Serve hot.

Tips

如用絞肉代替新鮮香腸，可自行加入調味後醃一會，便可繼續烹調。

To use minced meat instead of salsiccia, add seasoning to the meat and leave to marinate for a while before cooking

肉醬拌細寬蛋麵

Egg Tagliolini Bolognese

材料
Ingredients

細寬蛋麵500克	500 grams egg tagliolini
牛絞肉400克	400 grams ground beef
半肥半瘦豬絞肉200克	200 grams ground pork
洋葱1顆，切碎	1 onion, finely chopped
蒜頭2粒	2 garlic cloves
胡蘿蔔¼隻，西芹1根，切碎	¼ carrot, 1 celery rib, finely chopped
月桂葉3片	3 bay leaves
乾百里香1茶匙	1 teaspoon dried thyme
紅酒200毫升	200 ml red wine
原粒黑胡椒1茶匙	1 teaspoon black peppercorns
香草籽½條	½ vanilla pod
濃縮番茄醬2湯匙	2 tablespoons tomato paste
罐頭番茄1罐（400克）	1 can canned tomato (400 grams)
冰無鹽奶油30克	30 grams chilled unsalted butter
適量帕瑪森起士，磨碎	Grated Parmesan cheese to taste

做法
Method

1. 將4湯匙油的油溫加高，將牛絞肉及豬絞肉炒乾至微微焦香，盛起。
2. 於同一鍋內，爆香蒜頭，加入洋葱、胡蘿蔔及西芹一起炒香。
3. 將絞肉回鍋，加入月桂葉、百里香、黑胡椒粒及香草籽，再加入紅酒煮至半乾。
4. 加入罐頭番茄並壓碎，拌入番茄醬及足夠水分蓋過所有材料。煮滾後轉慢火煨煮1小時至絞肉熟透，調味。
5. 細寬蛋麵煮至有嚼勁，撈起拌入肉醬內一起煮至醬汁濃稠，最後拌入冰奶油後便可盛盤，佐以帕瑪森起士粉。

1. Heat 4 tablespoons oil and sauté ground beef and pork until brown and dry. Dish.
2. In the same pan, heat oil with garlic cloves, add onion and sauté until soft. Add carrots and celery and sauté until lightly caramelized.
3. Return the meat to the pan with bay leaves, thyme, black peppercorns and vanilla pod. Stir in red wine and reduce by half.
4. Add canned tomatoes, tomato paste and enough water to cover all ingredients. Bring to a boil and then simmer over low heat for 1 hour until the meat is tender. Adjust seasoning.
5. Cook egg tagliolini in salted boiling water until al dente. Drain and add to hot Bolognese sauce. Stir in chilled butter. Spoon onto serving platter and sprinkle with freshly grated Parmesan cheese.

Tips

1. 選購可馬上食用的千層麵皮，便可省去預先將麵皮煮熟的程序。
2. 使用此類麵皮需要較多醬汁才能將麵皮徹底烤透。烘烤時留意肉醬是否需要添加入適量水分。
3. 先將焗烤盤刷油，可防止千層麵黏著焗烤盤，較容易取出。

1. Get instant lasagne sheets that do not require pre-cooking.
2. More liquid sauce is needed in order to moisten the lasagne sheets properly. Add more liquid to the meat sauce if necessary.
3. To ensure the lasagne can be removed easily, brush the baking dish with oil before assembling the lasagne.

焗烤肉醬千層麵

Bolognese Meat Sauce Lasagne

材料
Ingredients

千層麵皮8片	8 pieces lasagne sheet
（大小18×8公分）	(size:18cm x 8cm)
肉醬汁1份	1 portion Bolognese meat sauce
白醬200毫升（參閱第115頁）	200 ml béhamel sauce (refer to p.115)
帕瑪森起士粉50克	50 grams grated Parmesan cheese
切碎的義大利巴西里1湯匙	1 tablespoon chopped Italian parsley

做法
Method

1. 將肉醬汁與半份白醬拌勻。
2. 焗烤盤內先排入兩片千層麵皮，塗上 ⅓ 份肉醬，再蓋上兩片千層麵皮，再塗上 ⅓ 份的肉醬及蓋上麵皮。重複再多做一層，最後蓋上麵皮。
3. 將剩下的白醬塗在麵皮上後，撒上起士粉。
4. 鋁箔紙封好，放入已預熱至180度烤箱烤30分鐘。拿開鋁箔紙後再烤至表層呈現金黃色。
5. 肉醬千層麵取出，待 15 分鐘後便可切成小份，撒上巴西里碎末後趁熱食用。

1. Mix Bolognese meat sauce with half of the béhamel sauce.
2. Line bottom of a baking dish with 2 lasagne sheets. Spread with ⅓ of the meat sauce. Cover with another 2 lasagne sheets and spread with more meat sauce. Repeat one more time and finish the top with 2 lasagne sheets.
3. Smother the remaining béhamel sauce on top and sprinkle with grated Parmesan cheese.
4. Cover with aluminum foil and bake in preheated oven at 180˚c for 30 minutes. Remove the foil and continue baking until a brown crust is formed on top.
5. Remove the lasagne from oven and leave to stand for 15 minutes. Cut into portions and sprinkle with chopped Italian parsley. Serve hot.

白醬雞肉釀義麵捲
佐新鮮番茄汁

Creamy Chicken Cannelloni with Sauce Vierge

材料
Ingredients

義大利麵捲10條	10 cannelloni
雞胸肉2片	2 chicken breasts
白醬200毫升（參閱第115頁）	200 ml béhamel sauce (refer to p.115)
清雞湯200毫升	200 ml chicken stock
肉豆蔻粉少許	A pinch of grated nutmeg

新鮮番茄醬汁
Souce vierge

熟番茄1顆	1 ripe tomato
羅勒葉8-10片，切細絲	8-10 leaves fresh basil, shredded
蒜頭1粒，拍碎	1 clove garlic, crushed
檸檬汁1湯匙	1 tablespoon lemon juice
初搾橄欖油50毫升	50 ml extra virgin olive oil

做法
Method

1. 雞胸肉用熱清雞湯燙熟，取出撕成細絲。
2. 雞肉與半份白醬拌勻，加入鹽、胡椒粉及肉豆蔻粉調味。
3. 義大利麵捲煮至微熟，取出將其中一面切開，釀入雞肉，並排放在已刷油的焗烤盤內，淋上其餘的白醬。
4. 封好鋁箔紙，放入已預熱至180度烤箱內烤30分鐘。
5. 另將番茄切小丁，與其餘材料拌勻並調味。義大利麵捲取出放在盤子上，佐番茄汁食用。

1. Mix Bolognese meat sauce with half of the Poach chicken breasts in chicken stock until cooked through; break into fine shreds.
2. Mix the chicken meat with half of the béhamel sauce and season with grated nutmeg, salt and pepper.
3. Boil the cannelloni until just soft. Remove and cut open on one side. Stuff with chicken meat and roll up. Arrange in a baking dish brushed with oil, seam side down. Top with remaining béhamel sauce.
4. Cover with aluminum foil and bake in preheated180˚c oven for 30 minutes.
5. To prepare sauce vierge, dice the tomatoes and mix with other ingredients. Season with salt and pepper. Serve with chicken cannelloni.

醬汁 的製作

Sauce Base Making

白醬
Béhamel Sauce

🥄 1.5 杯 cups   ⏰ 5-10 分鐘 mins

材料	Ingredients
鮮奶250毫升	250 ml milk
動物性鮮奶油100毫升	100 ml cream
可立即使用清雞湯50毫升	50 ml instant chicken stock
（可不用）	(optional)
奶油30克	30 grams butter
麵粉30克	30 grams flour

做法 | Method

1. 奶油煮融，拌入麵粉，邊攪拌邊用小火煮約2分鐘。
2. 慢慢拌入鮮奶、動物性鮮奶油及清雞湯，攪拌至滑順，繼續用小火煮至濃稠，期間需不停攪拌。
3. 白醬過篩至滑順，備用。

1. Melt butter in a saucepan, stir in flour and cook over low heat for 2 minutes. Keep stirring.
2. Stir in milk, cream and chicken stock gradually. Stir until smooth, keep cooking over low heat until thickened, stir occasionally.
3. Strain the béhamel sauce and keep for further use.

基本清雞湯
Basic Chicken Stock

 約1.5 公升 L　1 小時 hrs

材料	Ingredients
雞1隻，去皮並切成4塊	1 chicken, skin removed and cut into 4 quarters
胡蘿蔔1條，切大塊	1 carrot, cut into chunks
西芹2根，切大塊	2 stalks celery, cut into chunks
洋葱1顆，切塊	1 onion, cut into chunks
白胡椒粒1湯匙	1 tablespoon white peppercorns
乾百里香1茶匙	1 teaspoon dried thyme
乾月桂葉3片	3 bay leaves

做法　Method

1. 所有材料一起放湯鍋內，注入約2公升冷水。
2. 水滾後改用小火熬約1小時，不用蓋上蓋子。
3. 雞湯煮好後瀝掉湯底殘渣，放涼後可放冰箱或冷凍存放。

1. Place all ingredients in a soup pot and fill with cold tap water (about 2 litres).
2. Bring to a boil, then leave to simmer without lid for approx. 1 hour until flavourful.
3. Drain and leave to cool. Keep chilled or frozen.

濃雞高湯
Brown Chicken Stock

約 1
公升 L

2
小時 hrs

材料	Ingredients
雞1隻，去皮及切成4塊	1 chicken, skin removed and cut into 4 quarters
胡蘿蔔1根，切大塊	1 carrot, cut into chunks
西芹2根，切大塊	2 stalks celery, cut into chunks
洋蔥1顆，切塊	1 onion, cut into chunks
黑胡椒粒1湯匙	1 tablespoon black peppercorns
乾百里香1茶匙	1 teaspoon dried thyme
乾月桂葉3片	3 bay leaves

做法

1. 雞塊放入200℃烤箱烤至金黃，取出。
2. 將烤好的雞塊與其餘材料一起放湯鍋內，注入約2公升冷水蓋過所有材料。
3. 水開後用小火熬煮約2小時，不用蓋上蓋子。
4. 雞湯煮好後，瀝掉湯底殘渣，放涼後可放冰箱或冷凍存放。

Method

1. Roast chicken pieces in 200˚c oven until browned.
2. Place the roasted chicken pieces with other ingredients in a soup pot and filled with 2 litres cold tap water to cover all ingredients.
3. Bring to a boil, then leave to simmer without lid for approx. 2 hours.
4. Drain and leave to cool. Keep chilled or frozen.

魚高湯
Fish Stock

2 公升 L 45 分鐘 mins

材料	Ingredients
白肉魚800克，清洗乾淨	800 grams white fish, gutted and cleaned
白葡萄酒50毫升或 ½個檸檬	50 ml white wine or ½ lemon
洋葱1顆，切塊	1 onion, cut into chunks
胡蘿蔔1條，切塊	1 carrot, cut into chunks
西芹2根，切塊	2 stalks celery, cut into chunks
乾百里香1茶匙	1 teaspoon dried thyme
乾月桂葉3片	3 bay leaves
白胡椒粒1湯匙	1 tablespoon white peppercorns

做法 Method

1. 所有材料放入湯鍋內，注入2公升冷水至蓋過所有材料。
2. 水開後用小火煮約45分鐘至煮出味道。
3. 瀝掉湯底殘渣後待涼，保持冷藏或冰凍。

1. Place all ingredients in soup pot and filled with 2 litres cold tap water.
2. Bring to a boil, then let simmer for approx. 45 minutes until flavourful.
3. Drain and leave to cool. Keep chilled or frozen.

鮮蝦高湯
Shrimo Stock

約 1.5 公升 L　　1 小時 hrs

材料	Ingredients
原隻小蝦400克（約½斤）	400 grams shrimps in whole (approx ½ catty)
胡蘿蔔½根，切塊	½ carrot, diced
洋葱½顆，切塊	½ onion, diced
西芹1根，切塊	1 stalk celery, diced
白蘭地酒2湯匙	2 tablespoons brandy
月桂葉2片	2 bay leaves
濃縮番茄醬1湯匙	1 tablespoon tomato paste
乾百里香½茶匙	½ teaspoon dried thyme
黑胡椒粒½湯匙	½ tablespoon black peppercorns

做法 | **Method**

1. 小蝦洗淨瀝乾水分；將兩湯匙油的油溫加高，放入小蝦爆香。
2. 加入切好的蔬菜一起炒至微黃，拌入番茄醬炒香，灑入白蘭地酒。
3. 加入香葉、百里香及黑胡椒粒。注入冷水（約3公升）蓋過所有材料，水滾後用小火煮1小時。
4. 瀝掉湯底殘渣後待涼，可冷藏備用。

1. Rinse and dry the shrimps; heat 2 tablespoons oil and sauté the shrimps until dry.
2. Add vegetables and sauté until light browned. stir in tomato paste, sprinkle with brandy.
3. Add herbs and enough water (about 3 litres) to cover all ingredients. Bring to a boil and let simmer for 1 hour.
4. Drain and keep chilled or frozen until served.

番茄醬
Tomato Sauce (Salsa Pomodoro)

🥄 400 毫升 ml   ⏰ 20-25 分鐘 mins

材料	Ingredients
400克新鮮或罐頭番茄，切塊	400 grams fresh or canned tomatoes, chopped
洋蔥¼顆，切小丁	¼ onion, finely chopped
胡蘿蔔¼根，切小丁	¼ carrot, finely chopped
西芹½根，切小丁	½ celery rib, finely chopped
蒜頭2粒	2 cloves garlic, crushed
乾百里香½茶匙	½ teaspoon dried thyme
月桂葉2片	2 bay leaves
橄欖油2湯匙	2 tablespoons olive oil

做法

1. 將橄欖油油溫加高，爆香蒜頭，加入洋蔥、胡蘿蔔及西芹炒軟。
2. 拌入番茄，並加入百里香、月桂葉及200毫升水，煮滾後用小火煮至濃稠，待涼。
3. 將醬汁攪拌至滑順，瀝掉湯底殘渣後冷凍備用。

Method

1. Heat olive oil with crushed garlic. Add onion, carrot and celery, sauté until the vegetables are soft.
2. Add tomatoes with thyme, bay leaves and 200 ml water. Bring to a boil and then leave to simmer until thick. Remove, allow to cool.
3. Blend the sauce until smooth. Strain and keep refrigerated.

蔬菜高湯
Vegetable Broth

 1 公升 L ⏰ 40 分鐘 mins

材料	Ingredients
大胡蘿蔔1根，切粒	1 large carrot, diced
大洋蔥1顆，切粒	1 large onion, diced
西芹2根，切粒	2 stalks celery, diced
白胡椒粒1茶匙	1 teaspoon white peppercorns
乾百里香½茶匙	½ teaspoon dried thyme
乾月桂葉2片	2 bay leaves

做法 | Method

1. 將所有材料放湯鍋內，加入足夠冷水（約1.5公升）蓋過所有材料。
2. 水煮開後，轉小火繼續煮約30分鐘。瀝掉殘渣後冷藏備用。

1. Place all ingredients in a large saucepan and add enough cold water (about 1.5 litres) to cover all ingredients.
2. Bring to a boil and then allow to simmer for 30 minutes. Drain and reserve for further use.

常用材料
Common Ingredients

酸豆 Capers

酸豆在香港稱為水瓜子或續隨子，是一種地中海植物的花蕾。酸豆一般都是於採摘後用鹽或醋醃製，常見於義大利菜烹調，如沙拉、義麵或比薩等。它亦是經典義麵醬汁「Salsa Puttanesca」的主要材料。

如用鹽醃酸豆，要先將鹽分沖洗乾淨，再用清水浸泡最少2小時，期間要換水一至二次以將鹽味沖淡。如用醋醃的便無須浸泡，但注意醋醃酸豆一般酸味較重，故一般烹調用鹽醃的較適宜。

Capers are buds of a plant usually found in the Mediterranean area. The salted and pickled capers are widely used in Italian cuisine for making salad, pasta, pizza, etc. They are also a key ingredient in the classic pasta sauce "Salsa Puttanesca" .

To make use of salted capers, rinse to remove the salt and soak in clean water for at least 2 hours to reduce saltiness. It may also be necessary to change the water once during the soaking. There is no need to soak pickled capers, but be aware that they are usually quite sour in taste. Therefore, salted capers are more used for cooking.

鯷魚柳 Anchovy

鯷魚柳，又稱鳳尾魚，是一種常用於地中海烹調的細小海魚。鯷魚可作新鮮食用，又常用於鹽醃後再封存於油內。它是經典義麵「Puttanesca Pasta」內的一項主要材料，與任何海鮮配搭都很合適。

Anchovy is a kind of small salt water fish commonly consumed in the Mediterranean area. They can be eaten fresh or preserved by salting in brine, then packed in oil. Anchovy is a key ingredient in the making of the classic "Puttanesca Pasta". It also goes very well with seafood.

橄欖 Olives

橄欖多種植於地中海及北非等地帶，分別有綠橄欖及黑橄欖兩類品種。橄欖可經壓搾取出油分，亦可經醃製去除其苦澀味，用作食用。

Olives are mostly planted in Mediterranean and North African regions. They are available in black or green species. They can be pressed to produce oil, or be cured to remove the bitterness and make them more palatable.

牛肝菌 Dried Porcini

牛肝菌氣味獨特，與多樣材料如肉類、海鮮等都能搭配，故成為義大利烹調中一項不可或缺的材料。新鮮牛肝菌味道清新，常用作沙拉；而乾牛肝菌的味道則更香濃，無論用作小火燉煮、醬汁、義式燴飯或義麵烹調等，都會使菜餚味道更為提升。乾牛肝菌只需用冷水或熱水泡軟便可，浸泡後的水也可留作烹調之用。

Porcini mushroom is highly praised in Italian cuisine for its unique aroma and versatility in cooking. Fresh porcini mushrooms can be consumed raw and be made into a wonderful salad. When dried, they deliver a pungent aroma that can be used for stew, sauce, risotto, pasta sauce, etc. To prepare, allow steeping in cold or hot water until softened, reserving the liquid too for cooking if required.

罐裝番茄 Canned Tomato

罐裝番茄易於家中儲存，方便隨時使用。市面上可找到已去皮的整顆番茄，或已去皮及切塊的番茄。用時可連罐內的醬汁一併使用。

Canned tomatoes are convenient for storage at home. There are whole peeled tomatoes, peeled and chopped tomatoes available. The preserving juice can also be used for cooking.

番茄醬 Tomato Sauce

已製成的番茄醬常在超市看見。如果番茄醬是做成基本醬料，因需要再加入其他材料烹調，最好是選用原味的番茄醬。

Ready made tomato sauce is widely available in most many supermarkets. If the sauce is used as a base with other added ingredients, best choose tomato sauce in its original taste without other flavourings.

太陽番茄乾 Sun-dried Tomato

太陽番茄乾有油浸的，用時取出就可以。另也有乾製的，需要先浸泡變軟才可烹調食用。

Sun-dried tomatoes are available in oil preserved, which can be used immediately after being removed from jar; or in dried form, which needed to be soaked and moistened before using.

義大利麵粉 Flour (Farina Type"00")

義大利麵粉Farina Type「00」是小麥粉的一種，「00」的意思是指麵粉被磨至非常細緻，適用於製作義大利麵條、比薩薄餅，以及多種麵包甜點。如找不到此種麵粉，可用「多用途麵粉」代替。

Italian flour type "00" is wheat flour used for making fresh pasta, pizza or many kinds of baking and pastry goods. "00" means that the flour is very finely ground, thus giving it a talcum powder like texture. All purpose flour can be substituted if this kind of flour cannot be found.

硬小麥粉　Semola di Grano Duro Flour - durum Wheat Semolina Flour

　　硬小麥粉的蛋白質含量較一般麵粉高，而筋度亦較強。製作新鮮麵條時加入硬小麥粉可令麵糰更黏合，口感更富彈性及形狀更鮮明。

　　Most pasta is made with durum wheat flour, a kind of wheat high in protein and gluten. To make fresh pasta with a mixture of durum wheat and flour type "00" will make a dough sticks better and holds its shape better.

義大利培根　Pancetta

　　義大利培根是鹽醃豬肉，有的會經過煙燻處理，但一般只經過醃製處理。很多經典麵條醬汁都會用到義大利培根作主要材料，如培根蛋汁義麵、培根茄醬螺旋麵等。

　　Pancetta is salt cured pork either smoked or unsmoked. It is commonly used to flavour dishes, especially in pasta sauces, e.g. Spaghetti Carbonara, Fusilli all' Amatriciana, etc.

帕瑪森風乾火腿　Prosciutto - Parma Ham

　　帕瑪森火腿是豬腿以鹽生醃，再吊起風乾，最長至18個月時間。遠近馳名及高品質的風乾火腿多為義大利中部城市帕瑪森出產；另一東北部城市San Daniele也是有名的產區。風乾火腿可切成薄片作前菜食用，也可加入鮮奶油製成一道簡易的美味麵條醬汁。

　　Prosciutto - Parma Ham is salt cured raw ham, hung and aged for up to 18 months. Parma is a city in central Italy renowned for producing such ham. Another famous town for producing good quality prosciutto is San Daniele, a city in north-eastern Italy. Thinly sliced prosciutto is often served as an antipasto, or it can also be used to make a simple pasta sauce with cream.

香腸　Chorizo

香腸常見於西班牙、葡萄牙及拉丁美洲烹調。多用豬肉製成，亦會加入香料如紅椒粉，有時也會加入辣椒製成辣味腸。

Chorizo is commonly found in Spanish, Portuguese and Latin American cooking. It is mostly made with pork and seasoned with paprika, and sometimes chili for a hot chorizo.

帕瑪森起士　Parmigiano Reggiano-Parmesan Cheese

產自義大利中部的帕瑪森起士是以牛奶製成，經過凝脂、定形及鹽水浸泡過程後，帕瑪森起士要儲存於冷凍環境最少12個月至24個月不等的時間。在製作過程中剩餘的牛奶殘渣，傳統上的用途是給會被製成帕瑪森風乾火腿的豬隻吃的飼料。

Parmesan cheese is a delicacy from Parma, city in central Italy, made from cow's milk. After the milk is made into curd, shaped and brined, the cheese is left to age for at least 12 months and up to 24 months. Traditionally, remaining whey from curd making is used to feed pigs from which Parma ham is made.

牛奶起士　Asiago Cheese

正式的牛奶起士是產自義大利東北部的阿爾卑斯山區，有時也可代替帕瑪森起士用於義麵烹調。

"Official" Asiago cheese is produced from cow's milk in the alpine area in north-eastern Italy. Asiago is treated as interchangeable with Parmesan cheese in some cooking.

馬斯卡邦起士　Mascarpone

Mascarpone 起士是一種雙重或三重起士，脂肪含量約為65%-75%，是一種新鮮起士，需要冷藏，最為人熟悉是著名甜品Tiramisu的主要材料，此外也可用作烹調用途。

Mascarpone is a double or triple cream cheese, with a fat content of 65%-75%. It is a fresh cheese and must be refrigerated. Mascarpone is the main ingredient for making Tiramisu, the infamous Italian dessert. It can also be used in cooking.

優格 Yoghurt

優格是牛奶加入乳酸菌後發酵的酸奶,含有豐富蛋白質、鈣質及維他命B等,全脂優格的脂肪含量約是3%與牛奶相當,口感較濃稠滑順,故可代替鮮奶油是較為健康的選擇。

Yoghurt is a dairy product produced by bacterial fermentation of milk. It is nutritionally rich in protein, calcium and vitamin B, etc. The fat content of full fat yoghurt is similar to milk at 3%, but it is richer and smoother in taste, which makes it a light and nutritional alternative to cream.

瑞可塔起士 Ricotta Cheese

瑞可塔起士是新鮮起士的一種,是牛奶或羊奶為主的乳製品產物。由一般起士生產後留下的乳清,提煉而成,其蛋白質較奶類更容易被人吸收,它的顏色潔白,味道略甜,除了是甜點製作的用料之外,還可用作烹調菜式或作沙拉用途,如肉醬千層麵、比薩薄餅、義式餛飩餡料等,其保鮮期較短,需要冷藏。

Ricotta is a fresh cheese produced from whey, a liquid separated from curds making cow's or sheep's cheese. The protein in ricotta is more easily absorbed by human body than milk products. It has a cream colour and is slightly sweet in taste. Apart from being the ingredients for making desserts, ricotta can also be used in cooking or making salad, e.g. meat lasagna, pizza, filling ravioli, etc. Ricotta should be consumed quickly and it requires refrigeration.

希臘費塔起士 Feta Cheese

費塔起士源自希臘,傳統是用羊奶製作,製成後放鹽水內儲存6-8星期,顏色奶白,味道鹹香,是全球最為人所熟悉的起士種類之一,可作多種用途,如沙拉、比薩薄餅、餡餅用料等。

Originated from Greece, feta cheese is traditionally made with sheep's milk. After the cheese is made, it is aged in brine for 6-8 weeks. Feta cheese is cream in colour and has a savoury flavour. It is probably one of the world's most well-known cheeses. It can be used in various ways for making salad, pizza or filling for pie, etc.

大廚做的*Pasta*真美味

作　　者	傅季馨
發 行 人	程安琪
總 策 劃	程顯灝
執 行 企 劃	譽緻國際美學企業社、盧美娜
主　　編	譽緻國際美學企業社、莊旻
美　　編	譽緻國際美學企業社
封 面 設 計	洪瑞伯

出 版 者	橘子文化事業有限公司
總 代 理	三友圖書有限公司
地　　址	106 台北市安和路2段213號4樓
電　　話	（02）2377-4155
傳　　真	（02）2377-4355
E-mail	service@sanyau.com.tw
郵 政 劃 撥	5844889 三友圖書有限公司

總 經 銷	大和書報圖書股份有限公司
地　　址	新北市新莊區五工五路2號
電　　話	（02）8990-2588
傳　　真	（02）2299-7900

http://www.ju-zi.com.tw

初　　版	2014年01月
定　　價	新臺幣 298元
Ｉ Ｓ Ｂ Ｎ	978-986-6062-71-1(平裝)

本書由香港萬里機構出版有限公司
授權在台灣出版發行

國家圖書館出版品預行編目(CIP)

大廚做的pasta真美味 / 傅季馨作. -- 初版. -- 臺
北市：橘子文化, 2014.01
　　面；　　公分
ISBN 978-986-6062-71-1(平裝)

1.麵食食譜　2.義大利

427.38　　　　　　　　　　　　102026187